A Theory of Immediate Awareness

A Theory of Immediate Awareness

Self-Organization and Adaptation in Natural Intelligence

by

Myrna Estep
Indiana University,
Bloomington, Indiana, U.S.A.

Springer-Science+Business Media, B.V.

A C.I.P. Catalogue record for this book is available from the Library of Congress.

ISBN 978-90-481-6251-2 ISBN 978-94-017-0183-9 (eBook)
DOI 10.1007/978-94-017-0183-9

Printed on acid-free paper

Originally published by Kluwer Academic Publishers in 2003.
Softcover reprint of the hardcover 1st edition 2003

Dedication

This book is dedicated to the memory of my father, Modest S. Estep and my brother, James S. Estep

Contents

LIST OF FIGURES

PREFACE

This book is multi- and interdisciplinary in both scope and content. It draws upon philosophy, the neurosciences, psychology, computer science, and engineering in efforts to resolve fundamental issues about the nature of immediate awareness. Approximately the first half of the book is addressed to historical approaches to the question whether or not there is such a thing as immediate awareness, and if so, what it might be. This involves reviewing arguments that one way or another have been offered as answers to the question or ways of avoiding it. It also includes detailed discussions of some complex questions about the part immediate awareness plays in our over-all natural intelligence.

The second half of the book addresses intricate and complex issues involved in the computability of immediate awareness as it is found in simple, ordinary things human beings know how to do, as well as in some highly extraordinary things some know how to do. Over the past 2,500 years, human culture has discovered, created, and built very powerful tools for recognizing, classifying, and utilizing patterns found in the natural world. The most powerful of those tools is mathematics, the language of nature. The natural phenomenon of human knowing, of natural intelligence generally, is a very richly textured set of patterns that are highly complex, dynamic, self-organizing, and adaptive. We seek to understand those patterns by means of those powerful tools of mathematics and the aid of computers, looking for the most fundamental rules that bound or govern those patterns of natural intelligence. More specifically, we seek those fundamental principles of immediate awareness as it is exhibited in human know how.

Some readers may find some sections a bit laborious and difficult to follow. For that I apologize. Where you find it a bit rough going, please feel free to simply skip over those sections and try to pick up where your understanding takes you. For some of the more intractable concepts and arguments, I've tried to clear more than one pathway, providing many examples, to allow the journey toward understanding to continue.

For the sake of simplicity and to keep the book within manageable limits, there are issues I touch on here for purposes of clarifying the fundamental issues, but do not pursue to any great length. Included in these are issues related to indexicality in theories of language and theories of meaning acquisition. Most of these are more directly related to *knowledge that* issues that I largely set aside here so as to more fully focus upon the dynamics of natural intelligence as it is exhibited in *knowing how* and immediate awareness. However, I believe a fully developed theory of natural intelligence must develop theories showing the complex interrelations among all categories of knowing. There are other issues, for example mental representation theories that I also touch upon, but have not pursued. As the reader will quickly see, my primary focus here is upon experience that is present, not experience that is represented.

Myrna Estep

ACKNOWLEDGMENTS

This book grew out of a study that began many years ago while I was still a student at Indiana University in Bloomington. The natural beauty of the Indiana countryside and the I.U. campus seem quite inevitably to have led me to a life-long fascination with living things and theoretical attempts to model them. In this, I have been greatly inspired by my former mentors, George Maccia and Elizabeth Steiner, during long sessions at their home in the heavily wooded area of Lampkins Ridge Road near Bloomington. My efforts continued over the years, later in my own home in the Texas Hill Country northwest of San Antonio while teaching for a branch of the University of Texas, and in Africa at the University of Zimbabwe, as well as in other far reaches of the world as I traveled on behalf of the U.S. Government.

All along, I have had the very generous encouragement, support, and intellectual inspiration from my philosopher-scientist husband, Dr. Richard Schoenig. I owe my greatest debt to him. I am also endebted to many friends, too numerous to mention, including Hector Ñeri Castañeda, one of the most productive and creative philosophers of any century. With his very kind understanding and tolerant explanations of very complex subjects central to my arguments, he has been a real inspiration to me. I have also been encouraged and helped beyond measure by Professor Alwyn Scott of the Department of Mathematics at the University of Arizona and the Institute of Mathematical Modelling, Technical University of Denmark; Professor Robert Trappl of the Department of Medical Cybernetics and Artificial Intelligence at the University of Vienna; and Professor Gregg Rosenberg of the Computer Science and Artificial Intelligence Departments at the University of Georgia. These men have made enormous contributions to the field, and took time out of their heavy schedules to offer me valuable advice on ways to improve my efforts to clarify very complex ideas.

Of course, my sincerest gratitude is to my parents, Mary and Modest Estep, who always encouraged me as I was growing up, in spite of my often-spirited resistance. My mother will never know how much her own love of ideas and history has been an intellectual inspiration to me, and how much I have dearly loved our long conversations and debates. I regret that my father did not live to see the publication of this book. His respect for and love of the natural world formed much of the foundation for my own. There are many others who helped me, in one way or another, and to whom thanks are due. In particular, Dr. Elda Estep Franklin, Juanita Estep, Paul Estep, Dr. Dave Franklin, Betty Ann and Pat McGeehan and their entire family. I also thank S.A. David Shepard and other Federal LEOs in the San Antonio area whose names I do not know.

FIGURE ACKNOWLEDGEMENTS

I am grateful to the following for permission to reproduce illustration material:

Figures *FOUR-2* and *FOUR-3* reproduced (modified) with permission of Professor Titus Vilis of the Department of Physiology and Pharmacology, University of Western Ontario, London, Ontario, Canada.
Figure *SIX*-3 reproduced (modified) with permission of Pearson Education, Inc., Upper Saddle River, NJ, from *Neural Networks: A Comprehensive Foundation* by Simon Haykin, 1994.
Figure *SEVEN-1* reproduced (modified) from Wolfram Research, Inc., 2002.

INTRODUCTION

In the early 1970's, Hubert Dreyfus recognized a fundamentally serious problem with the Good Old Fashioned Artificial Intelligence (GOFAI) research program based on a top-down, logic approach to computer programming. He recognized a problem that in the last decade or so has been recognized by others who in turn shifted their efforts away from the top-down approach towards neural networks or connectionist research. Some of these research efforts are now known as Artificial Life. In sum, Dreyfus recognized that the top-down logic and knowledge based information-processing efforts on the classical Von Neumann digital computer are incapable of representing what he referred to as *commonsense know how* and understanding of human beings. He implicitly recognized that *knowing how* is not reducible to *knowledge that*,[1] the propositional knowledge representable in symbolic information-processing systems. He further recognized that the Cartesian rationalist model on which classical AI was based was doomed to failure. The Cartesian model was doomed because it is built on a fundamental split between the body and the mind.

Knowing how and commonsense understanding of human beings is highly complex, context sensitive, adaptive and dynamic knowing, involving human interests, values, feelings, motivations, and most importantly, bodily capacities and sensitivities, that go to make a human being.[2] *Knowing how* is far more fundamental in our intelligence than *knowledge that* because it is logically, epistemologically,[3] and temporally *prior* to our knowing propositional (knowledge *that*) statements. Our *knowing how* shows up virtually the day we are born, if not before,[4] and is present long before we ever become language speakers. Even then, we must *know how* to use language, *know how* to read and write, and we must *know how* to determine what is relevant in a context, *know how* to recognize and discriminate among the mouthings, vocalizations, tones, and signings of language tokens, as well as gestures or other movements around us. Moreover, we must *know how* to imagine alternative possibilities in meaning, sounds, and actions; we must *know how* to sense and make sense of others' bodily actions and discern movements, and be *immediately aware* of a very large number of fine distinctions with our capabilities to sense things around us and in us.

In part, we *know how* to do these things because of the immediate awareness[5] manifested in our *knowing how* found in the patterns of our actions, interactions and transactions with objects around us and in us. We may not even be able to say what those objects are. There is a sense in which our immediate awareness of such objects extends beyond our ability to talk about them. Our knowing extends beyond our language,[6] beyond our ability to classify those objects of our immediate awareness.

Furthermore, among other things, the immense domain of physical as well as abstract particulars present in our experience, including sounds, scents, images, a myriad of constantly changing unique particulars such as colors, shadows, surface edges, and forms, render it impossible to reduce this *knowing how* to *knowledge that.* Traditionally, philosophy has treated *knowledge that* [or simply 'know*ledge*'], as our knowing classes and those objects or things that can be classified by our use of language. We can label, name or describe those class objects or things in declarative statements or alphanumeric symbols. In philosophical jargon, the term 'knowledge' refers precisely to that class knowledge, to knowing which is propositional. This is just another way of saying that knowledge is what we know that can be recorded, spoken or written, in declarative sentences. On the other hand, the term 'knowing' refers to the broader class which includes not only *knowledge that* but also *knowing how* and *immediate awareness* embedded and sometimes hidden within the structures of *knowing how*.

But in the history of philosophy since Plato and Aristotle, there has always been a recognized tension between knowing which can be publicly represented in language, and knowing which cannot. Some have argued that immediate awareness is the foundation upon which all our other knowing rests; without it, we can never directly experience and know anything of reality. Others have argued that there is no such foundation, that there must always be some interface between reality and us. That interface, usually language, is how we represent reality to ourselves. Some have also argued that all we really know anything about is the interface itself, not the reality that may or may not be beyond it. Still others have argued that we don't really know anything about that either, that we don't really know anything at all. This line of argument has led to a tradition bereft of its own moorings.

Over the past three thousand years, kinds of knowing have been variously defined and classified in many ways, from speculative or theoretical and practical intelligence, to *basic* and *nonbasic* knowledge, and then *knowledge by acquaintance* and *knowledge by description.* These have been followed by various arguments purporting to show that the more basic, practical, or acquaintance knowing is always reducible to language about that knowing. That is, it has been and is still argued that practical intelligence (*knowing how*) is reducible to speculative or theoretical intelligence (*knowledge that*), and the basic is reducible to the nonbasic, to knowledge by description. At bottom, the claim is that all we really have is language about reality, not reality itself.

Since Descartes fundamentally split mind and body in 1637, in part to satisfy Church authorities, whatever human beings do with their bodies has been considered either not *really* a part of intelligence at all, or is not a part with which we should ideally be concerned. The prevailing theme has been that anything of real significance about what human beings know how to do in their bodily capacities is reducible to language propositions or prescriptions about it. Thus philosophers have tended to speak only of know*ledge* and not of the broader concept, know*ing*. *Knowledge that,* represented in language propositions, is commonly held to be the only kind of knowing. At minimum, it is argued, *knowledge that* best represents the highest of human being, the highest of human intelligence and reason. *Knowing how,* and immediate awareness embedded within it, have tended to get left out of the picture of intelligence altogether.

From a purely theoretical point of view, however, Gilbert Ryle's *Concept of Mind*, published in 1949, proved to be something of a watershed distinction. He fundamentally proved once and for all that the two kinds of knowing are not reducible to one another, that there are at least two altogether different kinds of knowing. His arguments showed that '*knowing how*' names a *different kind of intelligence* altogether from the traditionally recognized *knowledge that* intelligence. And *knowing how* is uniquely a part of the one who knows how, in a sense that *knowledge that* is not, though he did not have an adequate explanation for how it was uniquely a part of the one who knows how. With the notable exception of Gardner's works,[7] the full significance of Ryle's arguments has yet to be recognized in fields that study the nature of intelligence and those concerned with mapping natural intelligence into machines. At least one of the facts about human intelligence evident in those arguments is that intelligence is not a single thing to be measured by true and false, "paper and pencil" tests. We are creatures endowed with multiple intelligences that differ greatly from one another in very interesting ways and are interrelated in highly complex, dynamic ways we have yet to understand. We do not even minimally understand how, and by virtue of what, those multiple intelligences are bound together to form a unitary whole, intelligent being.

Minimally, *knowledge that* is largely a public matter because, in principle, it can to a large degree be manifested in public, alphanumeric symbolic language structures that are separate from the person who knows. Those language structures are available to anyone to publicly inspect. On the other hand, *knowing how* is somehow manifested *in the person.* It is manifested in, among other things, what they do, how they do it, and the manner, sensitivity, timing, resulting in a *seamless quality,* with which they do anything they know how to do. It is that seamless quality in the performance of knowing how that reveals the immediate awareness in the person who knows how.

Knowing how refers to simple things we know how to do such as knowing how to tie our shoes or ride a bicycle to far more complex things such as knowing how to play a viola, knowing how, when, where, and with what appropriate pressure to apply the brakes while driving your car, knowing how to prove theorems or discover new ones. No matter how many rules and prescriptions we write out to tell someone how to do any of these things, knowledge of those prescriptions and rules will never be sufficient for one to know how to do any of them. The crux of the differences between the two kinds of knowing is in the immediate awareness of the knower. *Knowing how* is not reducible to *knowledge that* because immediate awareness, what James and Russell[8] referred to as *knowledge by acquaintance*, is embedded and sometimes hidden within our natural intelligence of knowing how. When we know how to do something, and show that we know how by actually performing some task, that performance exhibits or points to our immediate awareness of many, perhaps uncountably many, things.

Sometimes, the non-reduction of *knowing how* to *knowledge that* means that we know more than we can say or write about our own knowing. Often, our knowing can only be exhibited or disclosed in our actual doing of something. We cannot simply know even true descriptions of *knowing how* and have it be so that we know how. No matter how many true descriptions you know about performing surgery, unless you are already a surgeon, all that knowledge will not turn you into one. The proof would have to be found in the performing.

When we speak of intelligence, it is best we speak of know*ing* rather than know*ledge*. Indeed, it is more likely that *knowledge* is reducible to *knowing how* rather than the other way around. If we are to map natural intelligence into computational processes of any kind, we must first know what natural intelligence is. A more adequate and comprehensive view of knowing may provide a more adequate and comprehensive concept of intelligence.

Moreover, though he did not make the distinction, Dreyfus recognized the need to distinguish between *rule-governed* know*ledge* and *rule-bound* know*ing*. Rule-governed *knowledge that* is what we call a recursively enumerable set. That means that it is a set of things we can count. It is a computable set, on the standard digital computer, a set of problems or function instances that can be defined over discrete, countable domains. As such, there are effective algorithms we can use to address those problems. Algorithms are sequential decision procedures to generate an answer to a problem. *Knowledge that* problems are those requiring "yes" or "no" decisions. On the other hand, rule-bound know*ing*, which includes *knowing how*, requires a different concept or approach to its computability altogether, if it is indeed computable at all.

Minimally, *knowing how* requires an approach that generates dynamic self-organizing patterns of interactions among very large numbers of components or elements in the way something is done. That is, a *knowing how* problem requires that we look at the dynamic patterns in the actual doing of something.

Computationally, *knowing how* would require a massively parallel and distributed approach. But any given instance of knowing how to do something may not be entirely computable at all due to immediate awareness embedded within it. Rule-bound knowing behavior may be regular, even predictable behavior to some extent because it is *bound* by a rule. But 'rule-boundedness' simply means that once we input a [for example, real or complex number] function into a computer, *we simply have to wait for the computer to show us* how it will map the points onto a real or complex number graph. That is, for example, we have to wait for the computer to show us which points will fall within a fixed circle on that graph, and which ones will fly to infinity. If it is computable, rule-bound knowing would be found in the dynamic patterns of the components of doing which fall within a fixed circle on a graph.

In a sense, the behavior generated by the computer *is* its own shortest description. It is its own algorithm. In the vocabulary of computation theory, such algorithms are said to be *incompressible*. There are no compressed overall law-like descriptions of the behavior obtained with such algorithms. That is, we do not have a compact overall description or algorithm of what the computer will do *prior* to its generating the behavior (shortest description) or algorithm that it in fact generates of rule-bound knowing.

I later argue that such incompressible algorithms may be used to some degree to characterize or simulate rule-bound knowing, but they will fail to do so completely. Rule-bound knowing is made up of those components or elements of a set which include kinds of primitive[9] epistemic [cognitive] relations and their terms constituting *immediate awareness* in intersection with *knowing how*. But we must beware of mistaking the symbol *of* something for *the thing* symbolized. We must not mistake a representation of something for the something represented. To do so is to fall into a fallacious trap, leading to kinds of fantasy, metaphorical theorizing.

To generate rule-boundedness as opposed to rule-governedness, we can input arbitrary reals [or complex numbers], iterate certain functions on those numbers, maintaining mathematical operations as primary rather than reducing all computations to bit operations, and simply watch the dynamics of the numbers unfold.[10] We can use the computer to try to understand the dynamic, self organizing patterns of *knowing how*. But we must not mistake the patterns the computer shows us for the knowing itself. Actual *knowing how* will have immediate awareness embedded within it whereas the simulation will not.

This massively parallel and distributed approach means that it may be possible to characterize and simulate *knowing how* by iterations of mathematical functions, generating discrete as well as continuous dynamical mappings. But *knowing how* is not rule-governed, where this means we have explicitly formulated the compressible, *overall* algorithm or rule for each step by step procedure of each detail of the knowing behavior to generate "yes" and "no" answers. "Yes" and "no" answers are responses to a *knowledge that* problem. They are not answers to a *knowing how* problem. If we want to know if someone knows how to do something, we ask them, tell them, or otherwise direct them to do it. We then watch to see if the patterns, timing, and sensitivity of their doing show that they know how. In a sense, in the computational approach to a *knowing how* problem, the computer takes over and proceeds to show us what it can and will do, what behavior it will generate. And we have to simply wait to see what it does.[11] The shortest way to predict or understand what a *knowing how* system will do is to watch what it does.

This rule-bound (but not rule-governed) behavior was never recognized by the traditional GOFAI research program because of its adherence to the top-down, sequential rather than parallel approach to computing. It was also never recognized because GOFAI research was directed to a different set of problems formed as questions requiring "yes" or "no" responses. In essence, GOFAI is directed to *knowledge that* problems that are clearly rule-governed. On the other hand, it may even be the case that the rules of *rule-bound* behavior *are not* formalizable or capable of being made explicit in a sequential algorithm. GOFAI researchers assumed that if behavior is regular [in some sense] then it is *rule-governed* and is therefore computable on the standard Von Neumann computer. They assumed that all that was needed is enough *knowledge that* engineering and writing enough explicit rules. They assumed *knowing how* is reducible to *knowledge that,* an assumption earlier proved false by Ryle. This explains their emphasis on logic and knowledge-based information processing in classical AI, and its subsequent failures with commonsense *know how* and understanding of human beings that Dreyfus recognized.

The futility of such AI efforts is now well recognized and documented, even by early proponents of AI,[12] though there remains a pervasive assumption and misconception that all knowing can somehow be reduced to and represented as *knowledge that*. One sees the latter assumption in certain of Penrose's comments[13] but also in efforts by Lenat[14] to reduce what he calls "commonsense knowledge" to *knowledge that* statements, even though such commonsense knowing entails *knowing how*. But there is no exhaustive list of sentences about running and jogging which will represent the immediate awareness of one who runs and jogs. Though there may be elements of commonsense know*ing* which can be so represented, such representations will not exhaust the category of commonsense knowing because there is that commonsense *know how* of which Dreyfus spoke.

The recognition that *knowing how*, including commonsense knowing and understanding, cannot be reduced to *knowledge that,* and the fact that characterizations of *knowing how* require a massively parallel and distributed approach to computing, has been supported by extensive research in the Artificial Life community. This recognition is also now generally accepted in the larger Artificial Intelligence community. Active research efforts are now directed to the *generation* of behavior, including generating the abilities of the computer to simulate life or intelligence in its *patterns* of behavior. The emphasis has changed from a logic and know*ledge*-based engineering program to a greater emphasis on mathematics and complex dynamic, self-organizing connectionist or neural network algorithms which *generate* not know*ledge* solutions to "yes" or "no" questions, but *knowing behavior* in the patterns of its simulations.

Following in the tradition of William James and Bertrand Russell, I give meaning to the concept 'immediate awareness' as a set of primitive relations of *knowing* made evident in what human beings know how to do. Unlike James and Russell, however, I argue for *nonpropositional* awareness and present my own theory that includes primitive relations of *touching* and *moving* of *knowing how*. This is intended to overcome the Cartesianism of earlier views, breaking down the fundamental conceptual split, found predominantly in Western cultures, between *knowing that* and *knowing how*. My extended theory of immediate awareness is presented within a broader theory of *sign* relations, not limited to alphanumeric *symbolic* relations that continue to dominate current theories of mind. It is a more complete classification than found in earlier theories.

I argue that primitive relations of immediate awareness are ultimately to be understood in terms of their information-theoretic properties, and a new distinction between what I call *rule-governed* and *rule-bound* knowing. This distinction is necessary to account for what Dreyfus referred to as commonsense know how, values, context sensitivity and understanding of human beings. I present a definition of immediate awareness which is at once mathematical, computational, epistemological and neurophysical, intended to pave the way toward resolution of fundamental problems in computational approaches to consciousness and our understanding of natural intelligence.

I also argue that the most promising approach to the computability of immediate awareness is a "weak AI" position involving the use of random Boolean networks and complex dynamical systems theory. My aim is to turn to a more *geometric* approach to immediate awareness and knowing how, as opposed to a *symbol-based* view. This is also an entirely new approach to the nature of immediate awareness not yet evident in the literature. In this approach, I follow some of the current research by Kauffman[15] and the use of random Boolean networks to exhibit fundamental properties of self-organization.
The use of random Boolean networks shows a way of obtaining *law-like properties* of those primitive relations of immediate awareness in terms of dynamical systems theory, without committing one to a physicalist/materialist theory. It gives us a way of understanding core properties of our own inner conscious lives, and of understanding the smooth and seamless sensitivity of primitive sensory and somatosensory-motor awareness. This direction for a theory of knowing [broader than just a theory of *knowledge about*] was implicit in the work of both James and Russell, though, among other things, they did not have the concept of nonlinear function, and Russell (as opposed to James) was wedded to an atomistic, summative ontology.

I also take issue with the standard "strong AI" position on the computability of immediate awareness by reassessing some of the very basic philosophic as well as computational and neurological concepts upon which our understanding of natural intelligence largely rests. For example, much of the current debate on the issue of consciousness is pervaded by prior Cartesian and reductionist assumptions which I critically assess. One of the most pervasive of these assumptions, already noted, is that all cognition generally is reducible to know*ledge that*, representable in declarative sentences (or encodable as such). One finds this assumption even in Penrose's work.[16] This neglects *knowing how* as a distinct and non-reducible kind of natural intelligence. It also neglects immediate awareness that is fundamentally embedded within *knowing how* structures, and which in turn is far more fundamental to our natural intelligence than *knowledge that*.

For ontological reasons which I present and argue for, I refer to immediate awareness as *knowing the unique*. The concept of uniqueness is based on arguments that objects of immediate awareness are sometimes *sui generis* objects, gotten by our use of non-logical, indexical operators found only in natural intelligence. *Sui generis* objects are unique, non-class objects of immediate awareness relations. I also show that there are basically three nonreducible kinds of human knowing or intelligence: propositional *knowledge that, knowing how,* and *knowing the unique* (immediate awareness). What Dreyfus earlier referred to as commonsense know how, values, context sensitivity and understanding of human beings is to be found in the intersection of knowing how and immediate awareness, our knowing the unique. That intersecting set includes kinds of cognitive relations, kinds of knowing, which are clearly not propositional or linguistic and not *rule-governed,* though I argue they are *rule-bound.* I also argue that this set is not decidable, hence not computable on the standard von Neumann computer.

Ultimately, the most complex, highly dynamic and intriguing facets of human natural intelligence are to be found in knowing how and immediate awareness. But we have barely scratched the surface of our own understanding.

---

[1]See Hubert Dreyfus, *What Computers Still Can't Do*, MIT Press, p. ix, 1992.

[2]Ibid.

[3]Epistemology is that branch of philosophy that addresses the nature of human knowledge, knowing, and belief. This includes an examination of the nature of evidence and justification for beliefs.

[4]Evidence for this has been around for decades. See T.G.R. Bower, "The Visual World of Infants," December, 1966, in *Perception: Mechanisms and Models*, San Francisco, W.H. Freeman and Company, 1972, pp. 349-357. Also see Peter W. Jusczyk, *The Discovery of Spoken Language*, Cambridge: MIT Press, 1997.

[5]The term 'immediate' [sometimes the term 'direct' is used] is not intended to mean meaningless, as Crick apparently assumes [see Crick, 1994, p. 33]. Note those philosophers who now want to know what kind of "thing" consciousness is.

[6]By 'language' I mean any alphanumeric symbolic written or spoken system, including unary, binary, denary.

[7] See Howard Gardner, *Frames of Mind: The Theory of Multiple Intelligences*, Basic Books, 1983.

[8] William James, *Essays in Radical Empiricism*, Harvard University Press, 1976 and Bertrand Russell, *Theory of Knowledge, The 1913 Manuscript*, edited by Elizabeth Ramsden Eames, Routledge, 1984. They called it *knowledge by acquaintance*, instead of *knowing* by acquaintance, in part because of the assumption that propositions are not necessarily tied to language.

[9] For the sake of readers who may not be familiar with this term in these contexts, the term 'primitive' basically means that "not derived from something else." A primitive object or relation is a basic object or relation that is not based upon anything else. In sound theories, both primitive and defined terms are used. The primitive terms are given meaning through the alternative terms; they are necessary to prevent circularity.

[10] There is a sense which I will explore to some degree later in this document, in which our *knowledge* [as representable in "that" clauses, definable over natural number domains or domains encodeable into natural number domains] may provide an index of the logical order of kinds of our knowing while *knowing how* [definable over real and complex number domains] is poised on the boundary between order and chaos.

[11] Stephen Wolfram, "Computer Software in Science and Mathematics," in *Scientific American*, September, 1984. Wolfram states the distinction in terms of *computational reducibility* and *computational irreducibility*. In computationally *irreducible* systems, general mathematical formulas [algorithms] that describe the overall behavior of such systems are not known and it is possible no such formulas can ever be found. For such systems, [and I argue that the intersecting set *knowing how and knowing the unique*, is such a system] we can only turn to explicit simulation of the behavior of that set in a computer. Computationally irreducible systems are not sets of computable problems that can be solved [with "yes" and "no" responses] in a finite time with definite algorithms as can *knowledge that* sets. Thus, as Wolfram points out and is a consequence of significance here, there are questions we can ask about the behavior of such systems that cannot be answered by any finite mathematical or computational process. Such questions are *undecidable.*

[12] The GOFAI top-down, sequential approach is not the appropriate approach to a *knowing how* problem, which is not a problem statable in sentences requiring "yes" or "no" responses. See Luc Steels, "The Artificial Life Roots of Artificial Intelligence, *Artificial Life*, Vol. 1, Number 1/2, MIT Press, Fall 1993.

[13] Roger Penrose, *Shadows of the Mind*, Oxford University Press, 1994. However, Penrose seems to be aware of the problem while not having the epistemological perspective and concepts to analyze it.

[14] See Lenat's own description of his enterprise in "Artificial Intelligence: A Critical Storehouse of Commonsense Knowledge is Now Taking Shape," in *Scientific American*, September, 1995, pp. 80-82.

[15] Stuart Kauffman, *The Origins of Order: Self-Organization and Selection in Evolution*, Oxford, 1993.

[16] Roger Penrose, *The Emperor's New Mind*, Oxford, 1984.

1. THE PROBLEM OF IMMEDIATE AWARENESS

"The pursuance of safe research will impoverish us all."[1]
Ian Stewart

This book addresses fundamental issues regarding the nature of the most intractable kind of consciousness called 'immediate awareness'. It also addresses the issue of whether or not immediate awareness, when it is found within the structures of *knowing how*, is computable [decidable] on the standard von Neumann computer. The term 'immediate' (sometimes the term 'direct' is used) is not intended to mean *meaningless* awareness, as some recent theorists assume.[2] Nor does immediate awareness mean "awareness *that,*" "conscious awareness *that*," or "consciousness *that*" such and such is the case, where a subject who is aware must know *that* they are aware in the sense of stating or otherwise indicating in language that they are aware. Nor does it mean that their awareness is necessarily accurate in some sense. Immediate awareness in the sense I am focusing upon here does not require that the one who is aware must be able to comment or reflect upon it or be right about it. Thus immediate awareness here also does not refer to "self-awareness," as some have recently defined the term 'consciousness' or 'conscious awareness'. Nonetheless, the sense of immediate awareness of concern here is *cognitive;* it is the most fundamental and pervasive faculty of human knowing underlying all natural intelligence. In a later chapter, I will present arguments providing a precise definition of immediate awareness, calling upon evidence from a variety of research studies.

The older realist philosophic tradition from which the term 'immediate awareness' comes emphasized a kind of "oneness" with the object(s) of one's awareness. In that tradition, "oneness" meant that the subject was not separated from the object of awareness with an intermediary proposition or sentence about it. There was no language interface between the subject and object. That is, one can be immediately aware of something, something is *meaningful*, according to this older tradition, even without language representations about that something. Immediate awareness was regarded in the tradition as a kind of experience, not necessarily limited to sensory experience, and theories about it ranged across kinds of realism, empiricism, pragmatism, and rationalism. Those early theories tended to rely upon introspective methods, and sometimes called immediate awareness "the given" to distinguish it from all other levels of experience. But "the given" is also called by some[3] a "blooming, buzzing, confusion" which is not what I am referring to. Moreover, the focus upon awareness in place of consciousness was in part an effort to avoid reification.[4] Awareness was not conceived as some "thing" occupying some place in the mind, but was a kind of relation between a person and reality. I will continue the older tradition of referring to awareness in place of the term 'consciousness' for some of the same reasons. And for reasons to be discussed with the reader in later sections, I will throughout refer to *immediate awareness* as *knowing the unique,* and will often use the terms interchangeably.

Problems surrounding the nature of *immediate awareness* can be found in the philosophical literature stemming back as far as Plato and are found in Descartes' *Meditations,* as well as in William James' and Bertrand Russell's writings on what they both called *knowledge by acquaintance.*[5] More recently, the term 'phenomenal consciousness' has been used in place of 'immediate awareness', though the two are not identical nor are they or equivalent in meaning, given certain assumptions about consciousness generally. Some recent philosophers[6] construe phenomenal consciousness, earlier called "immediate awareness," as nonintentional, which means that for them it is noncognitive; it is not a part of our knowing, it is not a part of our intelligence.

For arguments I make very clear throughout, I believe this construal is fundamentally mistaken. It is mistaken in part due to uncritical nominalist, idealist and behaviorist assumptions, as well as the uncritical uses of certain concepts. Without intending to get bogged down in philosophical jargon, we should perhaps clarify these concepts to some degree so that we may all see the overall direction that some have recently taken on the topic of immediate awareness.

1.1. The Influence of Nominalism, Idealism, and Behaviorism

Though there are variations and subtle nuances to be found in each of these doctrines, the essential properties of each are fairly clear. Nominalism basically holds that all abstract concepts and general terms (sometimes called "universals") have no objective reality "out there," independent of human beings. Anyone who holds a nominalist view is one who basically holds that all we have of what we may call "reality" is the *language* we use to describe it. All we actually have, they will claim, are language labels we use to name our experience. We have nothing of reality itself.

The pervasive influence of nominalism, combined with narrow behaviorism, based on an equally narrow empiricism, soon made it embarrassing to use terms such as 'immediate awareness'. With these influences, as lamented early on by James,[7] came the belief that everything must not only be represented or labeled with language to be known (or be an object of consciousness), everything can *only* be known through its label. Furthermore, *the only existing objects* are those for which we have a label. James called this the usurpation of metaphysics by language. It is also the usurpation of intelligence, of knowing, by language.

Accordingly, unless otherwise qualified in some way, a nominalist will deny that there is such a thing as immediate awareness since there can be nothing that is not mediated by language. We do not have any *immediate* contact with the world, the nominalist will claim. We can only have a *mediated* contact through language about our experience. Ultimately, all we actually have is our language. Over time, the influence of nominalism has led to wholesale confusion between symbols and the things symbolized, that in turn has led to collapsing levels of inquiry and fallacious inferences based upon the collapse. Perhaps the most pervasive fallacy of all is begging the question.

The doctrine of nominalism is closely associated with the doctrine of idealism, which basically says that reality is dependent upon the mind that cognizes it. There are no real objects "out there" independent of the minds that conceive and perceive them. And behaviorism, both scientific and philosophical behaviorism, along with varieties of naturalism and eliminative materialism, are doctrines that emphasize functional analysis and experimental empirical procedures, excluding all references to internal mental states, such as immediate awareness. Alternatively, they insist on reducing mentalistic expressions to descriptions of public behavioral, bodily processes.

Each of these doctrines has in common an undue emphasis upon language as the sole means by which we can have any cognitive relation whatsoever with the world. And even then, all we really have is language about our public experience.[8] They also have in common either an outright denial of a reality independent of human experience or the claim that even if it exists, we can't know anything about it. Moreover, they also deny the existence of internal, private states or they claim that if they exist we can't know anything about those either. All human beings can know—can have a cognitive relation with--is our language about our experience, a language that excludes terms referring to private, internal states. To allow otherwise is to fall into the Cartesian dualist trap or to engage in folk psychology.

Over many years, in one form or another, these doctrines have been so uncritically accepted, without an understanding of their historical antecedents and powerful arguments against them, that they are rarely questioned. The inherent errors within these doctrines have been compounded many times by various new movements, such as varieties of naturalism and the new cognitive sciences, advocated by their proponents to replace the efforts of earlier tradition. Moreover, the concepts 'consciousness' and 'awareness' are just two among a group of related terms, including the concept 'experience',[9] which have been disengaged from their traditional meanings as philosophers and some scientists pursue issues in consciousness studies, including issues about the nature of the mind and the nature of awareness.

Unfortunately, when a tradition is ignored, one risks "discovering the wheel" all over again. Though I will not dwell extensively on recent consciousness studies, with some few exceptions, I will tend to adhere to philosophic and scientific tradition. My reasons for referring to immediate awareness in place of consciousness will become clear later when I consider some neurophysical, psychological as well as ontological dimensions of the concept. In the process, it will also become evident to the reader that my own ontological position is one of an uncompromising realism. In its broadest sense, I believe there are real objects that exist independently of our experience, language and knowing of them, and these real objects have properties and enter into relations independently of the concepts with which we understand them or the language with which we describe them.

1.2. A Place for Ontological Questions

To clarify what I mean by the "ontological dimension of a concept," 'ontology' is a term sometimes used interchangeably with the term 'metaphysics'. But the latter has historically been associated with religious questions (*e.g.* the existence of God), that will play no part in my discussions. This religious association has been in spite of the original sense of metaphysics as "first philosophy," which meant studying the most general or necessary characteristics a thing must have to count as an entity or category of reality. 'Ontology' in the sense that I am using it, addresses itself to the nature of existing entities, such as individuals, properties, relations, and categories—fundamental concepts in any philosophic or scientific endeavor. It asks questions such as "What are the ultimate entities or things that exist in the world?"and whether or not some entities are reducible to others. In ancient Greece, Democritus [*c*.460-370 B.C.] was engaging in ontological questions when he asked and answered the question whether or not all material things are made up of atoms. But we can also ask ontological questions such as, "Is an individual existing thing reducible to a sum (list) of its properties?"

Over time, ontological questions have tended to get neglected given the increased influence of the above doctrines, especially the doctrine of nominalism. The rise of linguisticism and Postmodern literary influence in philosophy and the sciences have been consequences of the rise of nominalism and its emphasis upon language as the necessary (and only) interface we have with the world. Human cognition, or knowing in general, has almost become defined in terms of linguistic and literary categories and analysis, with a clear rejection of any immediate awareness of anything in the world, along with a rejection of the existence of anything independent of our language, including facts. Indeed, due to the influence of the above doctrines, the tradition reveals a tension in merely giving a name or label to the very concept of awareness itself as it is related to knowing and intelligence.

I believe the most promising way to address the nature of immediate awareness is by way of theory of knowledge (I prefer to call it with the broader concept *knowing*) and computability theory. One of the questions I address later is whether or not the intersecting set composed of the sets *knowing how* and immediate awareness (*knowing the unique)*, which I later refer to as *Boundary Set S*, is computable. To answer this question requires close examination of a broader theory and classification of knowing. That broader classification will include kinds of knowing not reducible to linguistically represented *knowledge that.* Both James and Russell, for example, recognized *knowledge by acquaintance* and they both claimed at various stages in their writings that it is not reducible to what James referred to as "knowledge about" and Russell referred to as "knowledge by description." Ryle later proved that they were right, referring to it as *knowledge that.*
It will also require a close examination of the *objects* of those kinds of knowing, especially from an ontological point of view. Following in the same tradition as James and Russell, I treat *knowing* as a kind of *relation* between a subject, S (one who knows), and an object, O, or objects (that which is known). The term 'object' must be understood in its broadest sense, to include even single features. It is a term in a relation.

My aim is to become clear on the different kinds of relations of knowing as well as the differences to be found among the objects of those relations. I am not retracing recent arguments found, for example, in Penrose's recent work,[10] though I believe he and I to some degree address the same kinds of problems. Penrose explicitly formulates the "strong AI" problem he is concerned to refute as a *knowledge that* problem,[11] even while his own arguments show the need for a broader classification, as well as a need for a broader understanding of the ontological *objects* of kinds of knowing. However, as many scholars today who are concerned with some of these same issues, Penrose does not explicitly address the ontological foundations of his inquiry at all.[12]

For purposes here, I tentatively understand *knowing how* as intelligent performances manifested or exhibited in patterns of doing. These intelligent patterns of doing are manifested in the manner, not style,[13] by which they are actually performed. Styles may be arbitrary, but as I define it, *manner* of performance (following extensive research on intelligent performances) is not. Contrary to popular conceptions, I hold that *knowing how* traverses the entire spectrum of all knowing, from bodily kinaesthetic performances such as tying your shoes, riding a bicycle, or playing a viola, to knowing how to perform purely abstract mental performances such as set theoretical operations, with the mind. We move and touch not just with our bodies, but also with our minds. Manner of performance will later be precisely defined in terms of timing and oscillation of movements, showing that it is indicative of immediate awareness and *knowing how*.

Moreover, manner of performance must be understood within a broader theory of indexicality, the means by which we point to, or disclose, our own knowing. We point to and disclose our own natural intelligence with both physical and abstract indexes when we do things we know how to do. The seamlessly timed and smooth manner by which we do something is an indicator of our knowing how, as is our use of abstract images of the mind. *Knowing how* is a kind of cognitive relation between a subject and objects. The objects of *knowing how* are patterns of performance in further relations with primitive epistemic elements of immediate awareness. More precise definitions and understanding of *knowing how* and immediate awareness, *knowing the unique*, will be forthcoming in later chapters, as well as extensive focus upon their highly complex interrelations.

The theory of immediate awareness presented here is built in part upon an extension of Russell's *knowledge by acquaintance.* In the realist tradition, his *knowledge by acquaintance* comes closest to being a complete theory of immediate awareness, though close analysis shows too much of a nominalist and Cartesian influence. Moreover, though his analysis uses outdated concepts of memory and the senses, including imagery, he nonetheless left us with a rigorous formal approach to what is perhaps the most complex and difficult subject around. I have chosen to make some modifications to his analysis of the primitive relations of acquaintance (awareness), and I make other changes as well which I will elaborate upon later.

My reasons for looking more closely at Russell rather than James in this regard will become clear as I proceed, though I believe Russell has gotten much wrong that James, to my mind, got right. Moreover, it will be clear to the reader that I fundamentally reject Russell's atomistic, summative ontology. In my view, immediate awareness, as the core of our total overall natural intelligence, is a highly complex, nonlinear, self-organizing and adaptive system. It is due to the intersection of *knowing how* and *immediate awareness* that natural intelligence systems exhibit complex, dynamic self-organization, adaptation, emergence, and continuous *transactions* [not merely *reactions* or *interactions*] with the environment, found only in living things.[14]

Nonetheless, Russell's rather uncompromising analytical and realist approach in the use of logical analysis has left us with a very useful classification of primitives on which his theory was built. With some changes, I retain many of the primitive epistemic relations of his *knowledge by acquaintance*, but extend beyond them to include other primitive relations he did not. I set forth what I argue is a more complete classification of kinds of primitive relations of immediate awareness so as ultimately to examine the laws and law-like properties of those relations, particularly in their relation to *knowing how*.

For purposes here, I will not directly address contemporary issues surrounding *knowledge that*, or awareness *that*. In particular, I will not address the philosophic definition of 'knowledge that' as "justified true belief" and the debates between those positions called *internalism*, *externalism*, and *reliablism*.[15] Indeed, I have endeavored to set *knowledge that* aside almost entirely, except where it is necessary to describe and explain its relation with immediate awareness and knowing how. Contrary to a prevailing trend, my effort will be to try to conduct an inquiry with a minimum of words ending in *ism*.

Moreover, I should make clear that this is not a neurophysiological study of the mind. However, I believe that anyone who has a serious interest in awareness can and should consult the tremendous research literature which can inform us about the brain and our neurological system generally. Where I believe clarification and evidence can be provided by referencing neurological studies, I do so.

1.3. Historical Background of the Problem: The Dualist Legacy of Descartes' Crooked Question

There is a well-known aphorism that basically states the following: "If we ask Nature crooked questions, we should not be surprised when she returns crooked answers."[16] Much of contemporary theory about human knowing and intelligence generally can be said to be based upon a crooked question put to Nature by Descartes. Descartes had posed the problem or question of mind as follows: Given that the principle of mechanical causation cannot tell us the difference between intelligent behavior and non-intelligent behavior, what *other* causal principle can tell us that difference?

The very logic or form of the question itself demanded a bifurcation, a "split," between body and mind. The answer left to us by Descartes is a dualist theory in which there are rigid laws explaining mechanical processes of the physical, as well as rigid laws explaining non-mechanical processes of the mental. He had posed a crooked question, and as Ryle [1949] pointed out, the answering theory was broken-backed as well from its inception. The answering theory still adhered to the grammar of mechanics in its efforts to claim the existence of both minds and bodies, and in its efforts to explain how the mind can influence the body.

Descartes' answering dualist theory assumed what Ryle later referred to as the "intellectualist legend" to explain the move from knowing a proposition, sentence, statement, or rule, to any rational action, practice, or performance. That is, practical reasoning is simply "to do a bit of theory, then do a bit of practice." That legend is mirrored in a prevailing bias in our conception of intelligence as solely involving "mental" or mind *knowledge that* with the use of alphanumeric natural or artificial languages.

On this view, the actual doings of performances or tasks with the body are commonly thought not to be directly associated with intelligence. Hence, from a cognitive or intelligence point of view, the actual doings of things are thought to be of less significance. The answering theory emphasized *knowing that* proposition, prescription or rule, as a necessary and sufficient condition to any rational action or rational performance. It has been a history of an emphasis on language, on propositions,[17] to the neglect of the facts of actual performance. Knowing how to do something, from knowing how to swim, ride a bicycle, fire an M40A1 on target at 300 meters, play chess, prove a theorem, or surgically probe an incision for a diseased organ, is explained largely in just those terms. But the "fit" of the explanation is not entirely comfortable.

The Cartesian theory of mind left us with what is known as the *Dogma of the Ghost in the Machine.* The Dogma states that "Minds are things but different sorts of things from bodies; mental processes are causes and effects, but different sorts of causes and effects from bodily movements." It was Descartes' theoretical effort to save freedom of the will by representing minds as *extra* centers of causal processes. Minds were held to be like machines but different from them. The now classical formulation of that Dogma stems from Descartes posing the wrong question to resolve difficulties of mind with the new Galilean mechanical [causal] view of the universe. The answering theory pressed what Ryle later showed to be a bundle of category mistakes into service to maintain the Dogma. The Dogma was the crooked answer to Descartes' crooked question.

Some contemporary corollaries of the Dogma and the crooked question that led to it are the equally hollow efforts to reduce mental states to the physical or reduce the physical to the mental. Both Idealism and monistic Materialism, the latter especially found in recent efforts of neuroscientists such as Crick and Koch [1992; 1994], are answers to improperly formed questions. Virtually all current hypotheses that I know of attempt to provide some explanation of awareness or consciousness generally by deriving awareness or consciousness from neural events in the active cerebral cortex. They end up with a kind of dualist mental and physical parallelism. Moreover, though contemporary neurophysical efforts are right to reject the "black box" view of the brain, their approach to the mental exhibits the same reductionist fallacy which is corollary to the classical Dogma. This fallacy was earlier recognized by Russell in his critique of James' theory of neutral monism:

> . . . if we are considering whether or how the sense of sight gives knowledge of physical objects, we must not assume that we know all about the retina, for the retina is a physical object of which we

> obtain knowledge by seeing it. Thus to assume that we know this or that about the retina is to [fallaciously] assume that we have already solved the *epistemological* problem of the physical knowledge to be derived by sight.[18]

In other words, the fallacy is to assume the facts of physics and physiology *as premises* upon which a theory of knowing, including immediate awareness, must build from the start. But the facts of physics and physiology are actually *inferences* from far more fundamental primitive awareness and experience. The fallacy is an instance of begging the question.

This same fallacy is evident in more recent cognitive science efforts, heavily influenced by nominalism and varieties of naturalism, to understand the link between our conceptual understanding of the world and our neurological system in terms of metaphor.[19] Even the hybrid concept of "schema" in Kant's sense of the term, is already an inference from more fundamental cognitive categories of awareness. Thus it assumes what it seeks to establish. Metaphor and schema cannot be assumed as premises from which a theory of awareness or consciousness must start.

This fallacy is also evident in the assumption made by some that our knowledge of the neurophysical system, at the classical or quantum-mechanical level, will solve the *epistemological* problem of our immediate awareness of objects, which we obtain (in part) from our senses of sight, touch, smell, taste, hearing. Russell's vision example to demonstrate the fallacy is particularly appropriate given recent emphasis upon the neurophysiology of that system as an explanation of mental awareness or consciousness.[20]

Put differently, the fallacy is the assumption that, because we use our brains to think, because there are neural networks [connections] in our brain when we think, we can therefore solve the *epistemological* problem of knowledge, knowing, and awareness by operating on the brain or otherwise scanning it to observe its neural processing. It is similar to the claim that we can find the explanation of a man's *values* by performing an autopsy on him. First of all, at this stage of neurophysiological research, and assuming our definitions are adequate, we are still unable to fully map human awareness to neurological correlates.[21] Secondly, even if we could obtain such a mapping, that map will still not provide an epistemological account of human awareness and knowing. An epistemological analysis of human knowing, as opposed to a neurophysical analysis, is directed to those conditions by which we actually distinguish intelligent from non-intelligent performances.

Descartes had mistaken the logic of his problem, thus posing the wrong question. In place of the above Cartesian question, still largely the question asked or implied by neurophysiologists with the attendant "double life" theory, perhaps Descartes should have asked "By what criteria do we actually distinguish intelligent from non-intelligent behavior?" This is the question later suggested by Ryle as a means for amending the Cartesian Error, and who (along with Russell) approached the answer by means of epistemological theory for a complete account of mind and consciousness. It is in part by means of formal logical and mathematical models that we can distinguish between intelligent and non-intelligent behavior, and between *knowing how* and knowing *that.* It is also by means of such models that we can characterize the more interesting and problematic kind of knowing, *immediate awareness.*

1.4. From the Linguistic Turn to the Cognitive Naturalistic Turn

There is a largely accepted, though I think somewhat mistaken, view that analytical philosophy was born of a Linguistic Turn establishing the study of language as the foundation of philosophy. The much earlier views of Gottlob Frege and Bertrand Russell were overturned sometime in the late 1950's and early 60's, when it was claimed that the fundamental issues in philosophy had to do with the cognitive part of the mind. The primacy of language inherited from the analytic tradition was largely replaced with what is called the Cognitive Turn, establishing the mind or cognition generally as the proper focus of philosophy. This view of analytic philosophy is not entirely correct (one has only to read the earlier works of both Frege and Russell to see that they were realists).[22] And one has only to look closely at the Cognitive Turn to see that it is in fact largely an extension of the

Linguistic Turn based on theories of *natural* language and mental representation theories. The Cognitive Turn was initially supposed to establish the central and fundamental issues in philosophy as not only logically independent from, but *logically prior to*, problems about language. But that is not what happened.

In the 1960's, Quine rejected all of Cartesian epistemology. He argued that the attempt to ground all knowledge on a foundation was futile, and proposed that epistemology be "naturalized," reduced to psychology. To be sure, there were and are good reasons to reject Cartesianism, not the least of which is the theoretical split between body and mind. And to varying degrees, Quine's proposal has largely been adopted. Realist foundationalist theories of human knowing have largely given way to neo-pragmatist, naturalist and coherence theories. The latter have in common the primary significance they give to language (or linguistic units generally), causal theories of mind within a monist-materialist and evolutionary framework, and the mind as a symbolic language representational system. All philosophic problems about the mind turn out be problems about language, or problems in some way *dependent upon* problems about language. But closely examined, these theories turn out to be an extension as well of the Cartesian "intellectualist legend," upholding the Dogma of the Ghost in the Machine and its corollaries. One way or another, they hold that all knowing of any kind, including all knowing how or practical intelligence, is reducible to *knowledge that*, and that intelligence is a single unitary thing found in a linguistically, symbolically represented "mind." Later, I will examine Quine's naturalist arguments since they are a paradigm of much of this movement.

1.5. The *Knowing That* and *Knowing How* Distinction: Manner of a Performance and Multiple Intelligences

In *The Concept of Mind* [1949] Ryle sought to refute this Cartesian intellectualist legend by showing that the exercise of intelligence or skills, *knowing how*, cannot be derived from nor is it reducible to knowledge *that*. He also argued that to judge that another knows how is to look beyond the performance to consider the abilities and propensities of which the performance was an actualization. He emphasized that it is the *manner* or the way the performance or task is actually done, which is indicative of the achievement of *knowing how*. One cannot know *how* by knowing *that*.

Following the publication of *The Concept of Mind*, considerable discussion and debate followed disputing his claim that *knowledge that* and *knowing how* were two distinct kinds of intelligence. Some argued that one could not assert a proposition without being able to say [or write] the sentence that carried the embedded proposition. Thus, without that doing, without that knowing how to say or write there could be no proposition to assert. As Maccia pointed out [1989], it was this operational perspective on knowing which led to devising "programmed learnings" to direct learning tasks to doings. As one among many, Mager [1975], for example, argued that cognitive intelligence is represented by overt behavior. Behavioral objectives were held to contain all meaningful (cognitive, nonambiguous) propositions. But what the behaviorists and others who agreed with this approach did not consider was the *logical* status of *knowing that* and *knowing how*.

Israel Scheffler [1965] is one of the few philosophers following Ryle who took seriously the cognitive as well as the logical distinctions between these two kinds of knowing. He demonstrated that the concept *knowledge that* has a different cognitive as well as logical status than *knowing how*. For example, "belief that" can be substituted in every sentence containing "knowing that," but such substitution is not admissible for "*knowing how.*" A substitution of "belief that" for "know that" did not violate linguistic use, but merely weakened the strength of the assertion. However, on the other hand, substitution of "belief how" was never in accordance with common use. Scheffler's arguments supported Ryle's claim that knowing how was not reducible to *knowledge that*, and he also showed that knowing how to do something does not entail belief. These are two different kinds of cognition, two different kinds of knowing, two different kinds of intelligence.

If their arguments are sound, it follows that we cannot justifiably define the entire domain of human knowledge [more appropriately, *knowing*] in terms of belief, as we currently do. Moreover, since the concept of intelligence largely follows definitions of 'knowledge', there were implications which followed for theories of intelligence as well. If we define the entire scope or domain of intelligence in terms of only one kind of knowledge (as we currently do), then it follows that the definition is too narrow. To do so is to limit the scope of the entire domain of intelligence, of human knowing, to *knowledge that,* defined in terms of [justified, true] belief, when it should be extended to include knowing how.

Again, Scheffler and others[23] argued that *knowing how* and *knowledge that* name distinct kinds of intelligence. They are distinct kinds of cognition. Their arguments also established that performances are cognitive, but not reducible to mere psychomotor activities. '*Knowing how*' is the name of a distinct kind of intelligence employed in action.

However, Maccia argued [1989] that although Scheffler and others presented logical and conceptual refinements on Ryle's characterization of the concept '*knowing how*', the refinements did not go far enough. There are conditions for *knowing how* defining the *manner* of a performance, which include its *smoothness* and *timing*, involving the sensory and somatosensory-motor systems. There are also conditions for *knowing how* which must include a condition for deciding whether a performance by one who knows how is observed by another who may or may not know how. With respect to the latter, for example:

> . .a non-swimmer might manage to keep his head above water, yet an observer could question that that person knows how to swim. In addition to having the capacity and the facilities for swimming, the person must perform as a swimmer does. Of course a non-swimmer does not perform [as a swimmer does]. He beats the water. If he is lucky, he stays afloat. Whatever happens is an accident. It is not a controlled activity.[24]

At least, the activity is not controlled as one who knows how to swim would control it. It is this sense of control that is a clear index to the person's knowing how to swim. It points to knowing how exhibited in the *manner* of a performance. That manner may be observed by someone who does not know how to swim, allowing the observer to decide whether or not the person swimming knows how.

But manner of a performance is also indicative of yet another kind of knowing found within *knowing how*. It is an indicator, index, or *sign* of the immediate awareness in the actual doing of a performance. The manner of a doing, meeting certain conditions, is a signature of *knowing how*. But it is also a signature of immediate awareness, of *knowing the unique*. The manner of someone's doing traverses the entire spectrum of performances, from physical doings with one's body to abstract doings with one's mind.

The timing and smoothness of the actions of one's doings is a crucial component in manner of performance. In F.C.S. Bartletts' [1958] study of thinking and skills acquisition, he observed that the movements of a novice are oscillatory. That is, the movements of a novice trying to perform are characterized by wavering and uncertainty as found in extreme oscillations [in a sense to be clarified later]. As a physical skill is mastered, one develops timing and a smooth coordination of touching and moving so that the oscillations are reduced and the actions flow continuously, seamlessly and smoothly.

Moreover, this same kind of oscillatory wavering and uncertainty can be found in novices learning to do kinds of abstract operations with the mind. Young students learning how to prove theorems for the first time evidence a "jaggedness" and hesitation in their performance clearly demonstrating their lack of familiarity and first hand experience not only with the rules of logic and math, but also with patterns of arguments, numbers, and functions.

In contrast to a novice, the actions of the one who knows how are smoothly connected. There are no spatial or temporal "gaps" or jarring, hesitant, wavering and uncertain actions in the patterns of the performance.[25] According to Maccia, the smoothness condition defining *manner* in *knowing how*, earlier recognized by Ryle, not only distinguishes *knowing how* from an accidental "happened to be," it also distinguishes a step-by-step procedure for doing something from an actual doing of that something. It is evident that one can know a step-by-step procedure for doing something without being able to actually do it. We can and do have experts in many kinds of *knowing how* who themselves are unable to perform, such as Olympic judges.

Moreover, over time and with experience[26] in the mastery of a skill, there is also an increased refinement in one's manner of doing something. There is an increased refinement in the immediate awareness of the elements, the particulars and patterns involved in the actual doing of a performance. Concurrently, there is an increased refinement as well in the immediate awareness of the particulars and patterns of the *context* or environment in which one performs. That is, the sensory and somatosensory systems of one who knows how become more "tuned" to the elements and patterns of a task or performance necessary to a smooth performance, and tuned as well to the surrounding elements and patterns within which one must perform to achieve a final objective, goal or terminus of the task.

A good swimmer, for example, will be immediately sensitive to the conditions of the water in which he or she swims, such as temperature, depth and force of water, as well as their effect on the swimmer's body, physical energy and psychological resources. Based upon that immediate awareness, the swimmer must make appropriately timed and smoothly patterned and coordinated directional movements with head, arms, legs, and torso while also adjusting one's periodic breathing before and after submerging one's head in the water. There are many possible patterns and combinations of patterns of movement in water, depending upon the water's conditions, including temperature, depth and force, upon the swimmer.

Moreover, swimmers report *imaging* themselves and their movements in their minds, depending upon their watery surroundings, as physically "slicing" the water and, with a variety of patterns of motion, "shoving" it behind them. How they mentally image their own bodies and physical movements as well as how they image the space within which they move their bodies, has an effect on the smoothness or manner of their actual swimming performance. Swimmers become highly immediately sensitive--with all their senses as well as their imagination-- to the conditions of the context in which they must perform, as well as conditions of their own bodies, internally and externally, their own psychological resources and movements. Thus, swimmers exhibit in their swimming the primitive relations of immediate awareness, including attending, sensing, imaging, (including *anticipatory* imaging), touching, and moving in their performances. They make their smooth patterns of movement appropriately or inappropriately, depending upon a great deal of immediate awareness and *knowing how*.

These same kinds of increased refinement in immediate awareness embedded within *knowing how* are found in the *knowing how* to do far more complex kinds of tasks or performances such as laparoscopic surgery. With recent advances in such surgical techniques requiring the use of probes equipped with cameras inserted into the body through small incisions,[27] medical schools have been forced to pay far more attention to kinds of primitive relations of immediate awareness embedded within *knowing how* to do this kind of surgery. Without that increased refinement or "tuning" in their actual performances, such surgical tasks can easily fail with disastrous consequences to a patient.

1.6. The Limits of Representation (Classification): The Role of Indexicals and Unique Objects Present

Crucial to my focus is a recognition that forming generalizations is only one part, perhaps ultimately a very minimal part, of our total intelligence. Forming generalizations is our capacity to classify and identify instances (objects) of classes, which includes our capacity for definition and partitioning. These capacities are a very important part of our *knowledge that* intelligence. But the entire domain of our intelligence cannot be limited to this one kind, because we are in fact endowed with multiple intelligences. That we are endowed with multiple intelligences is shown not only by Ryle's and others' arguments proving the nonreducibility of *knowing how* to *knowledge that* but also by the extensive empirical research by behavioral scientists and psychologists such as Gardner.[28]

This is not intended to undermine the significance of forming generalizations or *knowledge that.* But I have argued elsewhere[29] and continue to argue here that knowing unique individuals (particulars), or configurations of these, *unlike any other*, is more fundamental in the entire scope of human natural intelligence than forming generalizations. Moreover, immediate awareness, knowing uniques, is pivotally embedded in *knowing how*. To understand what a unique object or thing is, in comparison with any object of a classification of things, requires a prior understanding of the role of the indexical function in human intelligence.[30]

In general, an index is something we use to point to something else. We use the word 'indicator' or 'indexical' to mean the same thing, whether we are talking about something physical like a pencil that we may use to point to something else, or our physical gestures (like pointing with a finger or our head and shoulders), or something very abstract like a geometric form to "point" to a mathematical idea. An interest in the nature of indexes and their relationship to human language, thought and intelligence began in the 19th century when certain realists realized that we did not understand how this "pointing" function fit into our overall view of human language and reason. We later became aware that indexicality did not fit into the statement form necessary to standard first- and second-order logic. As far as I can tell, indexicality and its relation to natural intelligence, especially human knowing, has not been seriously considered by nominalists, idealists, or naturalist theorists.[31]

1.7. Analyze *This*

There is a well-known story of the Austrian philosopher Ludwig Wittgenstein and P. Sraffa, while riding on a train somewhere in Europe, were arguing over the "logical form and multiplicity" of language. During their argument, Sraffa made a gesture with his hand and fingers that meant something like disgust or contempt. And he asked Wittgenstein: "What is the logical form of *that?*" Sraffa was quite literally *pointing* to the limits of certain theories of language to account for the full scope of human thought.[32]

What is indexicality? In sum, indexicality is the structure and network of *thought* contents expressed or exhibited when we use indicators as signs to refer or point to items or objects of experience *as we experience them.* These indicators may be symbolic (linguistic), with such words as 'this' and 'that', what Russell sometimes called "logically proper names." But there are also *non-linguistic* indicators such as physical gestures, images in the mind, and patterns of performance or doing which may be *signs* or indicators of knowing relations between a subject and an object. They can be indexes of kinds of our intelligence. Our gestures, the use of images, and the patterns of our doing may be said to constitute a "signature" of our cognitive relations to objects. In a real sense, they *point* to our knowing, to our intelligence.

Indicators or indexicals, such as logically proper names (not limited to words such as 'this' and 'that'), as well as gestures, images, and patterns of performance, function indexically *to point to* objects of thought which may be unique objects of immediate awareness. When we use indicators to index or point to unique particulars or individuals of our experience *as experienced*, they are primitive indicators. When we do this, we are not classifying those objects. We are in fact selecting those objects for primitive levels of preattending or attention because those objects are *unlike any other*.

Indeed, it may be the case that there is no existing class or concept of a class by which we can sort such objects.[33] James early on pointed out the nominalist fallacy of assuming that humankind already has all the categories, classes or kinds it needs by which to classify any object *as experienced*. Conversely, he also pointed out the fallacy of also assuming that everything is an instance of some class. Primitive indicators point. They do not classify because they do not refer to classes or universals nor do they refer to instances of universals or classes. The objects of those indicators, which are particulars or individuals, are not class objects in part *because resemblance is not involved in our primitive selecting of them*. They are unique, *sui generis* objects, the objects of immediate awareness.

1.8. The Indexical Operator, Unlike Any Other: *Sui Generis* Objects

To understand what is being said here, we must understand the meaning of '*sui generis*'. We must also understand what it means to classify something. The phrase '*sui generis*' stems from the Latin '*sui*', which means "of its own" and '*generis*', which is the genitive form of '*genus*', meaning 'kind'. '*Sui generis*' is usually taken to mean "one of a kind, or a class of one." It is also sometimes defined, somewhat contradictorily, as "being the *only* example of its kind; *unique*."[34] However, 'kind' and 'class' are defined in terms of groups of entities that share certain properties in common. Those properties define membership in that kind or class. We classify things based on their properties and the *similarity* or *resemblance* of those properties with others of a kind or class. But when we select *sui generis* objects, we are selecting them as *like no other* or *unlike any other*, *in spite of* any property they may have in common with others. We are pointing to, selecting, a *sui generis* object as entirely unique, "of its own," because there is no kind or class of which *that one* is a part.

The etymology of the phrase '*sui generis*' reveals a tension. On the one hand, Indo-European languages claim that every thing is either a class, kind, or a member of a class or kind.[35] Hence the above definitions of '*sui generis*', especially "being the only example of its kind; unique." But there can be no *example* of a class or kind consisting of only one. The very meaning of 'example' requires a group of things of which *that one* is a part and is a member of that group because of properties it shares with others of the group. Yet that is not the root meaning of '*sui generis*'. Our Indo-European languages *also* acknowledge the existence of objects that are *not* classes, kinds, or members of either. Our Indo-European languages implicitly recognize objects that are entirely unique and cannot be classified as any kind. Nor do we "abstract" such objects from others of a kind or class, based on some rule of similarity or resemblance, precisely because they are of no kind or class. *Sui generis* objects are not sums or lists of their properties, as are classes and members of classes. And *like no other* is not a class operator, but a primitive indexical operator of immediate awareness.

This separating out of an object as unique is also not by identifying difference. Differences can be gotten from classification but uniqueness cannot. The concept *like no other* or *unlike any other,* which might be confused with the logical operator of *negation,* is in fact a nonlogical operator precisely because it is not an operator on classes or instances of classes. *Unlike any other* is an ostensive, *indexical* (or what is sometimes called "individuating") operator exhibited by signs. It is used in human thought to point to a *sui generis* object, individual, particular, or configurations of these. More will be said of indexicality, the use of proper names, and the indexical *like no other* to point to unique objects later. Let me summarize here by saying that when we select *sui generis* objects, the objects of immediate awareness, this is what I am referring to as the cognitive relation of *knowing the unique.*

Evidence will be presented in a later chapter showing that we do in fact select things as *sui generis.* Among other things, preattentive processing of information and the use of preattentive information for attention will show that this is the case. Furthermore, on all levels of the hierarchy of our primitive relations of immediate awareness, to "higher" levels of cognition, we continue to point to, select, some objects as unique. This can only be found in the relation between the subject and object(s); it cannot be found in the *descriptions* we give of those relations.

Aside from Russell's and some of Wittgenstein's[36] writings, most philosophic literature on the nature of indexicality focuses upon the role of linguistic indexicals, such as 'I', 'this', 'that', and 'now', within the context of language statements, propositions or knowledge by description, *knowledge that* statements.[37] That is, these efforts focus upon *word* meaning of indexicals, where indexicals refer or point to class objects. In general, they do not focus upon *speaker* (or more generally, *person*) meaning or awareness within primitive relations of immediate awareness, embedded within *knowing how*. This entails a broader very complex issue regarding what is called the *sense* and *no-sense* theories of proper names, an issue I will not pursue here.[38]

I will emphasize the role of indexicals where these are found in *knowing how*. This is another way of saying that *unlike any other* is not a concept of a *representation* relation, within linguistic, symbolic, propositional relations. In summary, the position I argue for is that the domain of our knowing is not exhausted by language acts of classification, definition, partitioning, by verification, and by falsification. Our knowing extends beyond know*ledge represented* within symbolic language structures and relations. It extends to knowing *sui generis* objects *present* in sign relations. One way we exhibit or manifest *knowing the unique* is by indexical acts within those sign relations, especially by our sensory and somatosensory *cum* motor relations found in *knowing how*.

1.9. The Basic Computational Idea and Argument

Artificial Intelligence research efforts have been misdirected to the extent that they have solely focused upon our capacity for classification, forming generalizations, and for operating with classes and their instances. They have also been misdirected to the extent that they have focused upon classification efforts in supervised or reinforcement learning. As such, AI research has simply focused upon rules of association, relevance, resemblance, or similarity to past events. In essence, computers are classification machines whether they are programmed with traditional top-down procedures or the more recent massively parallel, highly distributed neural network/connectionist procedures. By whatever language used for encoding, either unary, binary, or denary, these are all classification languages. Symbols are classes; 1's and 0's are classes; d^2, Euclidean distance to measure similarity of data presented, is a class; and all formulas are classes. In language that Russell may have used, computers are classification machines which handle *representations* of propositional *knowledge that*, but cannot handle *presentations* of immediate awareness, its primitive relations and objects.

It follows that the objects of classification machines must necessarily be class objects. On the view that all knowing is class knowledge (or *knowledge that*), all we can know are class objects that are sums or lists of their properties (or predicates). Whether explicitly or implicitly stated, the prevalent view is that all knowing proceeds by classifying objects according to some classification rule applied to the properties or predicates of the object. On this view, there are no such things as *sui generis*, unique objects, or if there are, we cannot know them.

However, although natural intelligence certainly classifies, unlike computers, classification is not all it does. We also cognitively select objects like no other in spite of properties those unique objects may have in common with others. Indeed, even neurophysical data shows that during what is called the "preattentive phase," before sensory perception of qualities in a stimulus actually occurs, the activity of neuronal groups in our brain are already being primed for that function. In other words, the activity of primitive selecting is already occurring in the absence of sensory (quality) perception. This primitive selection is part of immediate awareness. It is precisely the primitive relation of immediate awareness, involving *sui generis* objects, which is not computable or even capable of being simulated on a standard Von Neumann machine. And it is this primitive immediate awareness which Dreyfus implicitly recognized as the limit to AI.

As Dreyfus notes, "[A GOFAI system] . . .has no independent learning ability that allows it to recognize situations in which the rules it has been taught are inappropriate and to construct new rules. . .Networks. . .lack the ability to recognize situations in which what they have learned is inappropriate. . .what we really need is a system that learns [I would say *come to know*] *on its own* how to cope with the environment and modifies its own responses as the environment changes."[39] Supervised and reinforcement learning as presently construed cannot provide this; they cannot lead to a system which behaves appropriately in *unique* situations with unique objects because there are no such objects for these systems.

To understand the differences and relations between natural and artificial intelligence, what is needed, I argue, is a theoretical focus upon *immediate awareness, knowing the unique.* We need an exhaustive classification of the primitive relations and objects constituting the entire multi-layered, hierarchical array of categories of immediate awareness, including all the senses as well as the primitive relations and objects of moving and touching. In language Russell and James may have used, this is where objects are *present*, not where they are *represented.* Ultimately, we must also understand the highly complex interrelations between the multiplicity of these objects and our representations of the world. We especially need to focus on how immediate awareness is embedded within *knowing how,* since *knowing how* is largely the means by which immediate awareness is manifested in the public world.

Based on the above considerations, I will explore certain computational approaches to kinds of knowing found within the intersection of immediate awareness and knowing how, called *Boundary Set S*. I will discuss neural and random Boolean network conceptions of primitive knowing systems generally, and natural knowing systems in particular, as complex dynamical systems. Boundary Set *S* includes many kinds of knowing, including *commonsense know how*, such as knowing how to interrelate with other people and things in one's environment; and also knowing how to tie one's shoes, knowing how to drive one's car, knowing when, where, and with how much pressure to hit the brake, knowing how to prove a theorem, in general knowing the appropriate manner[40] with which to do a thing.

As noted earlier, I made the decision to focus here primarily upon knowing exhibited in *bodily kinaesthetic* tasks, but not to the exclusion of other kinds. Bodily kinaesthetic tasks involve highly complex primitive relations spanning the entire set of primitive relations of immediate awareness, including the primitive relations of *moving* and *touching*, where this intersects with *knowing how*. Analysis of just a few of those kinds of performances will clearly demonstrate this. Kinds of bodily kinaesthetic tasks include apparently simple ones (but difficult to actually perform) such as balancing a pin on its head, balancing on a tightrope, dancing, many athletic tasks and performances, to the more complex actions of a mime, conducting an orchestra or band, and the use of both physical and mechanical probes in minor to major surgical tasks. I will in particular consider a highly complex medical surgical task which, when properly analyzed, exhibits epistemic primitives of moving and touching, and patterns of performance found across the entire spectrum of that intersecting set, *Boundary Set S*.

It is precisely that high degree of complexity which will prove useful in providing a baseline characterization of *Boundary Set S*. Because of its complexity, if my analysis of the primitive relations found in tasks of this sort proves fruitful in the mathematical characterization of them, I will have effectively provided at least a *baseline idealization* of the entire set.

My theoretical computational strategy is in part to characterize knowing found in *Boundary Set S* as a multi-layer feedforward or recurrent self-organizing network. Because of the rather limited research in this area, I will also define mappings between the neural network of this knowing and random Boolean networks so as to utilize techniques already developed for studying random Boolean networks in their application to a neural network approach to *Boundary Set S*.

My strategy is based upon conceptualizing the dynamics of this knowing as it occurs in human beings *somewhat* along the lines of a model of self-regulating networks following much of the work of Kauffman[41] with random Boolean networks. But I am also following recent neural research into sensorimotor encoding at multiple levels of the somatosensory system.[42] My intention is to provide a partial answer to Dreyfus' call for a *hypothetical* theoretical mechanism to account for global sensitivity and the problem of contextual relevance in the performance of tasks. The model may be thought metaphorical, but it is not since our understanding of neural activity generally is based upon mathematical models of it. It is to those mathematical models that my model owes its origins.

Though many efforts to model *representations* of know*ledge* and adaptive learning, conceptualized as *knowledge that,* have surfaced, my focus of course is upon primitive categories of immediate awareness (the epistemic relation of *presentation).*[43] This is the most primitive consciousness by which we come to know the world. I view immediate awareness on the model of a massively parallel-distributed processing network of primitive relations and terms, that is, epistemic complexes. Those primitive relations or categories within immediate awareness, I argue, form the overall hidden layer of a network or population of neurons, conceptualized as relations, where data (particulars) from all parts of the input layer are combined at individual neurons. If neural networks are to be useful in generating characterizations of this kind of knowing behavior, that hidden layer itself must be conceived as a multi-layered hierarchy of primitive relations. Moreover, though I cannot thoroughly address the matter here, the units of this layer are found integrated throughout the other epistemological categories such as *knowledge that.* I will demonstrate the integration to some degree with *knowing how*, in an analysis of certain bodily kinaesthetic tasks, focusing especially upon the individual sensory and somatosensory-motor network of primitive relations of immediate awareness.

The basic idea is to conceive of that multi-layered hierarchy of primitive relations on a model of a recurrent unsupervised, *self-organizing* and adaptive multi-layer ensemble or network of neuronal structures manifesting synchronous activity. Then I examine that model for its adequacy as a model of primitive relations in Boundary Set *S*. The structure of the network of neurons exhibits distributed somatosensory and sensory topographic maps organized to respond to incoming particulars, as they become terms in the primitive relations. Coding of raw sensory and somatosensory input (data or particulars) occurs at the ensemble [network] rather than at a single unit [neuron] level and involves both spatial and temporal domains.

The hierarchy of primitive relations is conceptualized as arranged in "sheets" or layers where the visual, auditory, and somatosensory-motor relations are stacked in adjacent layers in such a way that particulars [terms] from corresponding points in space lie above each other. The effort overall is intended to follow the actual topological structure of the human brain[44] and multiple levels of the somatosensory system in mammals. This is so as to provide structural or architectural levels of organization in the neural network which I then characterize under learning and in terms of random Boolean nets.

The self-organizing neural network considered is intended to generate mappings from a high-dimensional signal space of raw sensory and somatosensory inputs, to lower dimensional topological structures. It does this by preserving neighborhood or similarity relations [Euclidean distance relations] in the input data, with the property to represent regions of high signal density on corresponding parts of the topological structure.

Overall, my argument strategy follows a somewhat typical *reductio ad absurdum.* That is, I start with the assumption that we can build a characterization of knowing behavior actually found in Boundary Set *S*. I will initially assume that we can use a set of sound algorithms, or a set constructed from them, for simulating kinds of knowing found there. Then I show a contradiction. I argue that existing computer neural network approaches are not sufficient to characterize or simulate the self-organizing and adaptive dynamics of knowing actually found in Boundary Set *S.* I also argue that the most promising approach to the computability and simulation of immediate awareness and *knowing how* is a "weak AI" position, involving the use of random Boolean networks and complex dynamical systems theory. This is so in order to turn to a more *geometric* approach to understand immediate awareness and *knowing how*, as opposed to a *symbol-based* view.

But I should make clear that I also believe there are fundamental limitations *in principle* to this approach. Along with Penrose, I believe that primitive immediate awareness of unique objects is not computable. Random Boolean networks can be used, however, to exhibit fundamental properties of self-organization of autocatalytic sets, and I argue that immediate awareness might be conceptualized as a kind of autocatalytic set of primitive epistemic relations. These are made publicly manifest in patterns of relations of *knowing how.*

What any theoretical effort aims for is an understanding of the most fundamental properties of phenomena, those properties which are invariant across all classes of the phenomena involved. Kauffman's use of random Boolean networks shows a way of obtaining *law-like properties* of those primitive relations of immediate awareness in terms of dynamical systems theory without committing one to a physicalist/materialist ontology and epistemology. This is in contrast to the approach to the same problem found in Penrose's quantum-mechanical theory of Objective Reduction, which I believe entails a fundamental materialist fallacy.

To some degree, I believe the approach with random Boolean networks, following Kauffman, is the most promising way toward a theoretical resolution of even the "hard" problem of consciousness. It may give us a way of understanding core properties of our own inner conscious lives, and of understanding the smooth and seamless sensitivity of primitive awareness. This direction for a theory of knowing was to some degree implicit in the work of both James and Russell, though they did not have the concept of nonlinear function, and Russell (as opposed to James) was wedded to an atomistic, summative ontology. Thus I map the neural network model onto random Boolean network structures, utilizing what we already know about Boolean networks to study the dynamics of primitive epistemic relational behavior of immediate awareness in an effort to derive law-like properties of those dynamics.

My effort also supports the view that *self-organization* and a phenomenon called *antichaos* are part of the development of natural intelligence. I refer to a 'knowing system', made up of epistemological sets, ultimately to be understood in terms of their information-theoretic properties. Epistemological sets encode information for making large numbers of epistemic elements, relations and objects. Different kinds of knowing differ in part because they have dissimilar patterns of epistemic *activity*, and because the elements themselves are different. Analysis of various kinds of bodily kinaesthetic tasks demonstrate the part played by the primitive elements of immediate awareness. This will also demonstrate the *emergent* properties of *knowing how* in dynamic interaction with one another, exhibiting self-organization, antichaos as well as chaos.

To clarify my use of certain terms, I am using the term 'emergent' in its scientific sense to refer to properties of *knowing how* which can, in principle, be understood from an analysis and understanding of the parts of *knowing how* plus a knowledge of how these parts nonlinearly interact.[45] Thus, I will seek ultimately to set forth the properties of the relations of immediate awareness as they are related to patterns of action of kinds of *knowing how*.

In summary, for the sake of my overall *reductio* strategy, a natural intelligence or natural knowing system such as a person, is conceptualized as a complex, dynamic and self-organizing, highly distributed, massively parallel-processing computer or network of relations. In those relations, epistemic elements regulate one another's activity either directly or through their products, that is *emergent properties,* of their interaction. The purpose of the model proposed is to understand the logic and mathematical structure of primitive epistemic regulatory activity of natural intelligence.

As I have made clear above, I take issue with the standard "strong AI" position on the computability of immediate awareness by reassessing some of the very basic *philosophic* as well as computational and neurological concepts upon which our understanding of intelligence [knowing] largely rests. But I also take issue with both the computational and neurophysiological research where models of classification processes of *concept* formation are often taken to be adequate to account for *percept* formation of immediate awareness, when they are not. Classification processes of computation cannot handle unique, *sui generis*, objects of immediate awareness. Additionally, all such computational models to date, as applied to immediate awareness, are either not self-organizing, while immediate awareness clearly is, or if they are self-organizing (as Kohonen's map) they are unable to accommodate the data streams and multi-layered, hierarchical nature of the sensory and somatosensory-motor structure of natural intelligence.

---

[1] Ian Stewart, "What Mathematics is For," in *Nature's Numbers: The Unreal Reality of Mathematical Imagination*, Basic Books, 1995, p.29.

[2]Francis Crick, *The Astonishing Hypothesis*, New York, Touchstone, 1994, p. 33.

[3] William James, *The Principles of Psychology, Volumes I and II,* 1890, London, Macmillan.

[4] Note those philosophers who now want to know what kind of "thing" consciousness is, such as Chalmers, *The Conscious Mind*, Oxford University Press, 1996; and Block, "On a Confusion About a Function of Consciousness, in *Behavioral and Brain Sciences*, Volume 18, 1995, pp. 227-287.

[5]There are very serious epistemological differences between James and Russell in their respective construals of *knowledge by acquaintance*. In essence, Russell's construal (at least in his 1913 theory of knowledge manuscript) permits nonpropositional immediate awareness whereas James' does not. See references to James, *Essays in Radical Empiricism*, Harvard University Press, 1976 and Russell's *Theory of Knowledge, The 1913 Manuscript*, Elizabeth Ramsden Eames, (ed.), Routledge, 1984.

[6]See Block, 1995. Block distinguishes between what he calls *phenomenal* consciousness and *access* consciousness, the latter representable in "that clauses". Though I agree that there is such a distinction to be made, I do not believe he has made it. Calling the more intractable kind of consciousness *phenomenal* already begs certain questions regarding the nature of the objects of that consciousness as well as the means of being conscious of them. The term 'phenomenal' refers to objects of the senses, that is things one is conscious of through the senses, as opposed to objects of thought or immediate awareness [or some, such as Penrose, say *intuition*]. Sorting the two [major] kinds of consciousness the way Block does may show obeisance to a prevalent nominalist *cum* empiricist tradition, but begging questions does not provide fundamental analysis.

[7] William James, "Some Omissions of Introspective Psychology," *Mind*, 9, January, 1884, pp. 1-26.

[8] The term 'public' here is intended to include whatever is operationally definable.

[9] Some recent writers on consciousness limit the term 'experience' to sense experience.

[10] Roger Penrose, *Shadows of the Mind*, Oxford University Press, 1994.

[11]See, for example, Penrose's descriptions of the problem he is addressing in *The Emperor's New Mind: Concerning Computers, Minds, and the Laws of Physics*, 1989, Oxford University Press, p. 10; also in relevant sections of his *Shadows of the Mind*, Oxford University Press, 1994, for example, p. 45. Penrose consistently confuses these kinds of knowing and recognizes only *knowledge that* problems with implied reductions of *knowing how* and *knowing the unique* to *knowledge that.*

[12]The absence of ontological analysis is noticeable, for example, in the works of Block, "On a Confusion About a Function of Consciousness, in *Behavioral and Brain Sciences*, Volume 18, pp. 227-287, 1995. Ontological analysis addresses the most fundamental kinds of things, objects, that exist.

[13]As will become clear, *manner* of a performance is not to be construed as *style* of performance. Styles may be arbitrary, but as I define it, *manner* of performance (following extensive research on intelligent performances) is not. Precisely defined in terms of timing and oscillation of movements, *manner* is indicative of *knowing how. Knowing how* is knowing where, when, what, in what way, and in what right proportion to do a thing.

[14]Though the position I argue for is essentially realist and contrary to traditional classical Cartesianism, I prefer to omit discussion of "isms" as far as possible in this study and simply attend to the inquiry at hand.

[15]See Louis P. Pojman, *What Can We Know, An Introduction to the Theory of Knowledge*, Belmont, Wadsworth Publishing Co., 1995. This work completely ignores *knowing how.*

[16] Peter Geach, *Mental Acts: Their Content and their Objects*, New York, Humanities Press, 1957. Geach referenced this aphorism as follows: "No experiment can either justify or straighten out a confusion of thought; if we are in a muddle when we design an experiment, it is only to be expected that [if] we should ask Nature cross questions . . .she return crooked answers."

[17]If one does not accept the existence of propositions, then the emphasis is on sentences or statements.

[18] Bertrand Russell, *Theory of Knowledge: The 1913 Manuscript*, 1984, p. 51[emphasis is mine]. For a statement of James' theory of neutral monism, see William James, *Essays in Radical Empiricism*, Longmans, 1912, especially the essay, "Does 'Consciousness' Exist?". Epistemology is that branch of philosophy that addresses the nature of human knowledge, knowing, and belief. This includes an examination of the nature of evidence and justification for beliefs.

[19] See George Lakoff and Mark Johnson, *Philosophy in the Flesh: The Embodied Mind and Its Challenge to Western Thought*, Basic Books, 1998. Also see Bernard J. Baars, *A Cognitive Theory of Consciousness*, Oxford University Press, 1998.

[20] See F. Crick, and C. Koch, "Towards a Neurobiological Theory of Consciousness," *Seminars in the Neurosciences*, 1990, Vol. 2, pp. 263-275 and "The Problem of Consciousness," in *Scientific American*, Volume 267, number 110, 1992.

[21] As an example of the effort to map the binding problem to neurological correlates, see: Andreas K. Engel and Wolf Singer, "Temporal binding and the neural correlates of sensory awareness" in *Trends in Cognitive Sciences*, Vol. 5, no. 1, 2001, pp.16-25. Also see: Chris Frith, Richard Perry and Erik Lumer, "The neural correlates of conscious experience: an experimental framework," *Trends in Cognitive Sciences*, Vol. 3, no. 3, 1999, pp. 105-114.

[22] This view is also based on an uncritical acceptance of what is called the Gradualist Thesis regarding language, originally stated by Quine [1951] in the "Two Dogmas of Empiricism." Gradualism basically states that there is no clear demarcation between formal (constructed) languages and natural languages. For many good reasons, there are powerful arguments against this thesis, some of which I touch upon in the following chapters.

[23] This thesis has been substantially supported by empirical and theoretical research on the nature of intelligence by Howard Gardner and others associated with the Harvard Project Zero Multiple Intelligence Theory. See Gardner references.

[24] See George Maccia, 1989 [my emphasis].

[25] Timing is inextricably a part of any intentional doing which is manifested in temporal sequences, as *knowing how* clearly is. Representing this computationally presents serious problems. See Jeffrey Elman, 1990.

[26] Frederic C. S. Bartlett, *Thinking*, New York, Basic Books, 1958.

[27] See Gary Stix, "Boot Camp for Surgeons," in *Scientific American*, September 1995, p. 24.

[28] See Howard Gardner, *The Mind's New Science*, 1985; *Frames of Mind: The Theory of Multiple Intelligences*, Basic Books, 1993.

[29] See especially M. Estep, 1984.

[30] To fully cover the subject of indexicality would require an entire book of its own. Due to the complexity of the subject, I cannot address the indexical function [within sign relations] as thoroughly and completely in this work as the subject warrants. However, see my 1993a and the Castañeda references to indexicality. Contemporary writers on the subject of consciousness, such as Crick, confuse the concept *sign* with the concept *symbol*, thus reinforcing faulty representational theories of the mind. I use the concept *sign* to refer to the category of all indexicals, more or less following Charles Sanders Peirce, *The Collected Papers of Charles Sanders Peirce*, Vols. I-VI, Charles Hartshorne and Paul Weiss, eds., Cambridge, Massachusetts, 1958. Thus, signs or indexicals include symbols, ikons (or images), and actions, including performances. This is necessary so as to theoretically capture the broader scope of all knowing, including all signs which disclose our knowing, and that which may be presented as well as represented.

[31] However, some interesting work on gesture recognition in the design of computer software is underway which I will address later in this book.

[32] Norman Malcolm, *Ludwig Wittgenstein: A Memoir,* Oxford: Oxford University Press, 1958, pp. 57-58.

[33] William James, "Some Omissions of Introspective Psychology," *Mind*, 9, January, 1884, pp. 1-26. Also note the distinction between selecting and sorting objects. To select does not imply the existence of a class; to sort does imply the existence of a class of objects.

[34] See the *American Heritage College Dictionary*, Third Edition, Boston, New York, Houghton Mifflin Company, 1993, emphasis mine.

[35] For an interesting article touching on this subject, see Alan Hausman and Tom Foster, "Is Everything a Class?" in *Philosophical Studies*, Vol. 32, 1977, pp. 371-376.

[36] See Ludwig Wittgenstein, *Tractatus Logico-Philosophicus*, London, Routledge & Kegan Paul Ltd., 1961.

[37] See David Kaplan, "Demonstratives," in *Themes From Kaplan*, Joseph Almog, John Perry, Howard Wettstein, eds, New York, Oxford University Press, 1989. Also John Perry's "The Problem of the Essential Indexical," in *NOÛS*, 13, 1979, pp. 3-21.

[38] See the *Appendix* on Proper Names.

[39] See the Introduction to the MIT Edition of *What Computers Still Can't Do*, pp. xxxviii-xxxix, emphasis mine.

[40] Again, *manner* of a performance is not to be construed as *style* of performance.

[41] Stuart Kauffman, "Antichaos and Adaptation," in *Scientific American*, Volume 265, No. 2, August, 1991; and *Origins of Order: Self-Organization and Selection in Evolution*, Oxford University Press, 1993. Also see the latest research underway in DNA computing by Leonard Adleman and Richard J. Lipton reported in "A Boom in Plans For DNA Computing" and "DNA Solution for Hard Computational Problems," *Science*, Vol. 268, 28 April 1995, pp. 498-499 and pp. 542-545 respectively.

[42] Miguel A. Nicolelis, Luiz A. Baccala, Rick C.S. Lin, John K. Chapin, "Sensorimotor Encoding by Synchronous Neural Ensemble Activity at Multiple Levels of the Somatosensory System," *Science*, Volume 268, 2 June, 1995, pp. 1353-1358.

[43] The terms 'representation' and 'presentation' are labels for two different kinds of cognitive or epistemic relation between a subject, S, and an object, O. To readers unfamiliar with philosophical and psychological terms, the use of more than one term or label to refer to the same thing may tend to be confusing. I will try to clarify as I proceed. In general, however, 'representation' refers to the "knowledge *that*" (language) relation while 'presentation' refers to an immediate (or direct) relation with an object. I call the latter "knowing the unique" or "immediate awareness."

[44] See: Kandel, E.R. and J.H. Schwartz (1991). *Principles of Neural Science*, 3rd edition, New York: Elsevier.

[45] Emergent properties are those resulting from nonlinear interactions among elements of systems. Very generally, a nonlinear system is one whose elements are not linked together in a linear or proportional manner. That is, the elements are not summative as are the elements in linear systems. Linear systems can be characterized by equations having the following form: $\varphi = x + y$. Nonlinear systems are those in which such an equation does not hold, that is $\varphi \neq x + y$.

2. THE PRIMITIVE RELATIONS OF KNOWLEDGE BY ACQUAINTANCE

"The words or the language, as they are written or spoken, do not seem to play any role in my mechanism of thought." Albert Einstein[1]

In this chapter, I focus on certain very fundamental issues related to any theory of whatever it is that allows us to become directly aware of something. In doing so, I start with what is perhaps the best realist analysis historically available to us, Russell's theory of knowledge by acquaintance, though I include considerations from James' theory as well. Given the prevalence of views which either deny that there is anything specifically "mental" about mental events, or deny that there are such things at all, there are a lot of good reasons to start with a realist who took their existence quite seriously. Russell's intention to avoid fallacies of all kinds, including that of assuming as premises the very facts one intends to prove or disprove, renders his analysis very useful. He has provided us with a carefully thought out, albeit faulty classification of many of the most basic primitive relations of what he called "knowledge by acquaintance." This was his concept of immediate awareness, though he did not clearly distinguish levels of awareness generally.

2.1. A Realist Theory of Immediate Awareness

In the philosophical literature, aside from Descartes' *Meditations*, Hume's *A Treatise of Human Nature*, and James'[2] various works, Russell left us with one of the more extensive and serious treatments of the subject of immediate awareness to be found. Let me start, however, by noting some of what's wrong with it. Coming out of the analytic tradition, one of the problems with Russell's *knowledge by acquaintance* is that it suffers from the *knowledge that* bias left over from Descartes' dualism. A consequence of this is he omits a number of primitive relations of immediate awareness, including *moving* and *touching* exhibited in bodily capacities and sensitivities of *knowing how*. Of course, his knowledge of the senses and overall somatosensory system, as well as memory was limited, given what was known at the time. He also had a very truncated view of what an image is and the part images play in our overall intelligence.[3]

A comprehensive and complete theory of immediate awareness certainly requires primitive relations of moving and touching that are not conceived to be purely motor or mechanical concepts (as Descartes and to some degree Russell held them to be), but are basic, fundamental parts of our overall multiple intelligences. Furthermore, again contrary to Cartesian dualism, not only do we move and touch with our bodies, we also in a real sense *move among* and *touch* abstract structures with our minds, using imagery as we do so. This occurs not just in fantasizing or daydreaming, but in things we do every day. Moreover, the imagery we use is not necessarily based on resemblance to any physical thing. We do not have to go far to find examples of this. Any college freshman who has used Venn diagrams to represent general ideas in a proof has had something of the experience.

Furthermore, although earlier philosophers, including Russell, studied the senses and their relation to human knowledge, they were largely unaware that the senses are part of a *single* larger neurological system called the *somatosensory-motor* system. Our understanding of this larger system and the rather complex interrelations among the component parts of it is relatively recent. That research shows that we cannot adequately address the senses and their relation to our knowing, specifically our *knowing how*, separate and apart from the larger somato- (body) and motor system of which they are a part.[4] Moreover, if one looks closely at the analyses of philosophers and some behavioral scientists who focus upon the senses, one finds both explicit as well as implicit references to bodily capacities and movements in those analyses.[5] This is so in spite of efforts to keep those references out of the analyses, reflecting in part a tension resulting from the Cartesian bifurcation between body and mind.

Moreover, among other things, Russell's analysis of knowledge by acquaintance, as complex and fully developed as it is, does not provide us with an exhaustive or adequate analysis of primitive relations between a subject, S, and an object(s) O. Of course, neurological and other research continues to this day to try to find all those primitives for the senses. We still do not have an exhaustive and adequate analysis even for the primitives of our visual system. But many of the inadequacies of Russell's analysis may be due to his sense data approach, atomism, and his Cartesian commitments. They are certainly not adequate to address commonsense *know how*. His Cartesian commitments, for example, led him to essentially ignore *knowing how* as a kind of knowing at all, a fault we still generally share with him. Related to this are his inadequate treatments of the concepts *experience* and *memory*, but also the relations of *sensation* and *imagination.*

We will see, for example, that his analysis of the primitive concept *experience* is a largely static concept, much the same as one finds in Descartes' *Meditations.* It not only provides no way to account for knowing how to do things with one's body, it also provides no way to account for the cumulative effects of experience (especially kinaesthetic bodily experience) upon human knowing over time. That is, Russell's knowledge by acquaintance does not conceptually provide the theoretical means to account for incremental growth and dynamics in human knowing and understanding. His very method of analysis, a sense datum *cum* additive view of phenomena, may have been the reason for this. He left us with a largely dormant, unmoving, undynamic concept of experience within which all of acquaintance or immediate awareness is found.

Related to this is his equally static and Cartesian "mind-centered" notions of the primitive relations or "species" of acquaintance or immediate awareness, *attention*, *immediate memory*, *sensation* and *imagination*. His account of immediate memory includes the use of nonlinguistic objects such as images, but his analysis of the *use* of images is largely tied to *knowledge by description*, that is their use in classification. As noted, he had no real concept of sensory and somatosensory-motor memory as we have today and so he could not have a concept of the part images play in that memory.

Our concept of memory, for example, includes a physical concept of neural memory based on empirical research in the neurosciences, such as the well-known Hebbian learning rule in neural network theory. There is a significant sense in which our physical sensory and somatosensory-motor systems, such as our senses of bodily touch, [including both sensory and the somatosensory system] moving, smell, and taste "remember" and learn from experience over time. Our senses, the primitive relations of sensation, the somatosensory-motor system, can become more refined or "tuned" over time.

Additionally, due to his sense data and additive view, Russell left out entirely any consideration of the use of "unreal" images in sensory and somatosensory-motor accounts of relations of *moving* and *touching*, required for our understanding of kinds of *knowing how.* Such "unreal" imaging is part of the use of probes in various kinds in anticipatory or exploratory knowing behavior found, for example, in *knowing how* to probe in surgical tasks, especially prior to the invention of laparoscopy. He held (very much reminiscent of Hume) that images are no more than "copies of past sensations."[6] Remnants of this view survive in the theoretical literature today in the reliance upon a principle of resemblance by which a child purportedly "abstracts" rules such as those allowing the natural numbers to go on to infinity.[7] Though Russell may have gotten much right in his analysis of experience, particularly his notion of *present* experience (acquaintance or awareness), there is much that is wrong and much that is missing.

Because of his implicit acceptance of what Ryle referred to as the "intellectualist legend", (that knowing how to do something is simply to know some rule or prescription then putting the rule or prescription into practice), and his static concept of experience, as well as his "mind-centered" notion of memory, Russell also did not recognize the problem of the limits to *knowledge that* rule-governedness. For Russell, as well as many other philosophers in the analytic tradition, knowing how to get from point A to point B is simply to know the rule or proposition which states that one takes so many steps from A in one direction, so many in another, until one gets to B. Among other problems with this view, to know a prescription or rule is not necessarily to know how to apply it. Knowing that one must take *x* number of steps to the right is not necessarily to know how to walk in the first place, let alone to know how to spatially orient oneself in an environment, or to know how to use one's sensory and somatosensory-motor system of primitive relations to move appropriately in the right direction.

With one exception in his theory of knowledge by acquaintance, he had no concept and provided no account of the relation between the primitive relations of knowledge by acquaintance to the bodily *manner* of actual performance. That is, he drew no connection between those primitive relations and the *actual doing* of something, specifically to the manner of actual moving and touching, indicative of *knowing how*. The exception to this is found in his treatment of the experience of time and the nature of acquaintance, immediate awareness, involved in our knowledge of relations themselves.

At an even more fundamental level, the analytic tradition's almost total reliance upon an atomistic, summative ontology, clearly evident in Russell's theory, does not permit the kind of analysis necessary to a comprehensive theory of intelligence. The kinds of self-organizing dynamics found in natural intelligence cannot be accounted for on a nondynamic, linear and additive model of kinds of knowing as found in the analytic tradition generally. A comprehensive account of natural intelligence requires a concept of *constitutive* or *configural* uniqueness and nonlinear dynamics which cannot be forthcoming from Russell's concept of *summative whole,* a requirement of his atomism and sense data approach.

2.2. Analysis of Experience: Russell's Knowledge by Acquaintance

To understand the scope of the domain of experience, we must first understand Russell's distinction between *knowledge by acquaintance* and *knowledge by description*. We should keep in mind the following broad distinctions used to sort the two.

Knowledge by Acquaintance (KA) is the most primitive and most pervasive aspect of experience, but is not identical to it. It is the experience of facts, particulars, and universals and is the direct *immediate* cognitive relation with an object. KA is also non-propositional in the sense that it is that experience which is entirely nonlinguistic, though it includes the use of what he calls 'proper names'. These should not be confused with names in language, such as 'Mary'. In his theory of knowledge by acquaintance, for Russell, a proper name is a *primitive* indexical; it is anything that may be an object of *thought* even if we are not directly attending to it. But he also uses it to refer to linguistic indexicals such as the words 'this' and 'that'. Moreover, knowledge by acquaintance cannot be communicated to another person by language description, and it cannot be in error. We cannot be mistaken about what we know by acquaintance. Furthermore, belief not a condition of KA. All cognitive relations of any kind, such as attention, sensation, memory, imagination, believing, disbelieving, *presuppose* acquaintance, but they are not identical to it. KA consists of knowing *primitive* facts in the world.

Knowledge by Description (KD), on the other hand, is not primitive and it is narrower in scope than is knowledge by acquaintance. KD is *mediated* experience of universals and is indirect experience with an object. It is also entirely propositional in the sense that whatever is known by description can be set forth in "that-clauses" (declarative sentences) of language.

Thus, knowledge by acquaintance (KA) is broader in scope than knowledge by description (KD), and consists of experience of facts, particulars and universals. In drawing this distinction, Russell was also emphatically trying to draw the distinction between the object-level of an inquiry and descriptions or language *about* that object. There are differences between the levels. Acutely conscious of fallacies of inference, Russell was keenly aware of the dangers of collapsing levels of inquiry, the danger of assuming that what one says about one level also applies to another.[8]

Fundamentally, in Russell's theory, experience is a relation between a subject, S, and object(s), O. The term 'object' is also taken by him in its broadest sense, to include features such as the color red. A *fact* is defined as the kind of thing that is expressed by the phrase "that so-and-so is the case," it is the kind of object towards which we have a belief, expressed in a proposition.[9] But for Russell, we can have immediate experience of facts. A *particular* is an entity which can only enter into complexes (a mix of terms and the relations that unite them), as the subject of a predicate or as one of the terms of a relation. It can never be a predicate or relation itself.[10] The concept *universal* includes primitive relations and predicates.[11]

The term *experience* is primitive, or undefined, but that does not mean it has no meaning. Russell “unpacks” and analyzes its meaning by considering alternative terms which he uses to explain it.[12] An analysis of the concept *experience* shows that at any given moment, a person is "aware" of certain things. He then narrows his focus from the concept *experience* in general, left primitive or undefined, to a related undefined term, *awareness.* In turn, *awareness* is equivalent with the concept *acquaintance* or *present* experience (whatever is present to us at any moment in time).

Acquaintance or awareness is further delineated in terms of what he calls its "species" of primitive relations. Each species of acquaintance he tentatively defines in terms of the primitive concept of experience, thus most of our attention will focus upon these. His method is a way of elucidating the meaning of a group of related primitive terms, which, as the term *primitive* suggests, are left undefined. Initially, we should understand Russell's efforts to explain the broader concept *experience* and its relation to the concept *awareness.*

Awareness [acquaintance, the relation of *present* experience] is a related primitive concept but may be understood as follows: If one is asked what one is aware of, one can reply that he or she is aware of *this* or *that.* That is, with the use of linguistic indexicals, the primitive proper names 'this' and 'that',[13] *if one is asked*, one can point to the object(s) of one's experience or awareness. Though he is using a hypothetical language example, the objects of immediate awareness are *primitive* objects. They are pointed to but cannot be described in language. Russell defines them as objects that are capable, in principle, of being called by indexically-functioning primitive proper names. Though this may be confusing, it is important to stress the hypothetical nature of his method of explicating the meaning of these primitive notions. He is using *prima facie* linguistic proper names in a hypothetical manner by which to elucidate how *non-linguistic, no-sense* proper names may be used in acquaintance, in immediate awareness. They are used to "point" to objects of that primitive immediate awareness.

It becomes clear that he holds that *knowledge by acquaintance* is not reducible to *knowledge by description.* The objects of acquaintance or awareness are not propositional and cannot be found in language descriptions at all.[14] The proper names 'this' and 'that' have an indexical pointing or ostensive function for Russell, and these are the only means by which we can understand the meaning of primitive objects. As primitive proper names in immediate awareness or present experience, they are *not* elliptical for definite descriptions or rigid designators.[15] That is, the meaning of the primitive concept *awareness,* and all the "species" or relations of awareness such as *attending, sensing, remembering, imagining,*[16] and so on, can be made clear with the indexicals or proper names, 'this' and 'that'. But their meaning cannot be communicated in language descriptions. This appears to be Russell's effort to distinguish between "awareness of" and "awareness that," where the latter may involve a subject's pointing (with language indexicals) about their awareness, though he does not explicitly state this distinction.

Moreover, according to Russell, so long as those names actually name and are not just meaningless sounds, the person cannot be wrong. The person's *description* of those objects of awareness, however, may very well be wrong. His position is clearly that one cannot, with certainty, publicly communicate with language descriptions to another person the objects of which one is immediately aware or acquainted. That is, one cannot assert public propositions or statements describing those things of which he or she is immediately aware. One can only indirectly make public one's knowing the objects of immediate awareness by indexical means.

However, there is a sense in which if one speaks to oneself, one can *denote* those objects of immediate awareness to oneself by indexically functioning proper names rather than descriptions. Again, one cannot be in error as long as the names actually name things, the objects of immediate awareness [which are primitive objects], at that moment of awareness. Due to confusions regarding the nature of error, hallucination, illusion, and the primitive relation of acquaintance, it is important to note that for Russell error is tied solely to representations or descriptions of objects. Error is not found in primitive immediate awareness or acquaintance.

Moreover, the object of a language representation, of knowledge by description (KD), is a *belief* and can be wrong. But the object of acquaintance or awareness may be a fact, particular, or universal itself, out there "in the world." A belief [of a language proposition] either corresponds to an independently existing fact out there in the world, or it does not so correspond. Reflecting his realism, Russell defines "truth" as follows: If a belief does correspond to a fact, then it is true; if it does not so correspond, then it is false. However, the object of acquaintance is a *primitive* object present or given in experience. It is not a belief. Belief is not assimilated to knowledge by acquaintance in Russell's theory, though it is assimilated to representations (knowledge by description).

Even if one has an hallucination, denoting [indexing] the object which is an hallucination to oneself, that object present in experience is not an error. In general these are the reasons he did not assimilate belief to knowledge by acquaintance. He wanted to prevent confusions regarding what one can be in error about.

In summary, awareness is a unified assemblage or complex of objects to which one can in principle give primitive proper names, and of which one cannot be in error, though one can be in error in one's descriptions of those objects.[17]

2.2.1. The Scope of the Domain of Experience

To determine the scope of experience, he asks the following questions: (1) Are faint and peripheral sensations included in the meaning of 'experience'? (2) Are all or any of our present true beliefs included in present experience? (3) Do we now experience past things which we remember? (4) What leads us to believe that "our" total experience is not the only experience in the entire universe? How do we know that our experience is not all-encompassing, the only experience in the world?

To answer (1) "Are faint and peripheral sensations included in 'experience'?" he focuses upon the field of vision.

> "Normally, if we are attending to anything seen, it is to what is in the centre of the field that we attend, but we can, by an effort of will, attend to what is in the margin. It is obvious that, when we do so,

> what we attend to is indubitably experienced. Thus the question we have to consider is whether attention constitutes experience, or whether things not attended to are also experienced."[18]

Now this is precisely a question which has recently recurred in debates on the nature and kinds of consciousness,[19] though those engaging in the debate appear to limit the scope of experience to *sense* experience [erroneously defined as 'phenomenal consciousness']. For Russell, the domain of experience is very large and includes objects the mind *does not* select. The domain of experience includes but is not reducible to primitive relations of acquaintance with objects.

The most primitive relation of the "species" of knowledge by acquaintance is attention. It is the selecting of *one* object, the *this* of our acquaintance, of our present experience. The *this* is a particular. Primitive attention is a selection among objects [particulars] that are before the mind of which one is aware. Hence it logically follows that there exists *a larger field of objects* from which the mind selects. That larger field of objects is the domain of experience. Because there is a larger field of objects from which attention selects, there are objects *not attended to* in the domain of experience from which our mind selects. These may include objects for which we have no name or label and may never have a name or label.

Primitive attention is the "starting point" or means by which one enters the "circle of cognition" for Russell. His concept is still largely in accordance with contemporary uses of the term, though in some ways, his analysis is more perspicuous because he recognized the indexical function found in awareness. Though for Russell, attention is the primitive point at which one enters the circle of cognition, the primitive selection of an object, many contemporary psychological theories, Maccia [1987, 1989], Cherry [1957], and others[20] use recognition as the faculty or principle of primitive cognitive selection.

To answer (2) "Are all or any of our present true beliefs included in present experience?" Russell states that our mental life is largely [though not exclusively] composed of beliefs and of what we call "knowledge" of "facts." A fact is:

> . . .the kind of thing that is expressed by the phrase "that so-and-so is the case." A "fact" in this sense [as part of a language description] is something different from an existing sensible thing; it is the kind of

> object towards which we have a belief, expressed in a proposition. . .*the question is, whether the facts towards which beliefs are directed are ever experienced.* It is obvious at once that most of the facts which we consider to be within our knowledge are not experienced.[21]

Facts within our knowledge, but beyond the scope of our experience, include the fact that London has six million inhabitants or that Napolean was defeated at Waterloo. He could also have included any number of other facts that we do not experience, but are nonetheless part of our knowledge, such as that there is a center point of the Earth or the galaxy; there is an infinite number of primes; and so on. There are also primitive facts, however, which we immediately experience for ourselves without relying upon the testimony of others or even upon our own reasoning from other facts. These are facts which form part of our present experience and which figure significantly in Russell's total theory of knowledge, and in his concept *presence*.

In answer to (3) "Do we now 'experience' past things which we remember?" Russell sorts out *intellectual* memory and *sensational* memory. When one knows that "I saw Jones yesterday", this is intellectual memory and is one of those primitive facts mentioned above. With respect to sensational memory, we must not confuse the objects of true memory and the present images of past things. "I may call up now before my mind an image of a man I saw yesterday; the image is not in the past, and I certainly experience it as now, but the image itself is not memory."[22] But there is an immediate memory of something, for example, which just happened in which the thing that just happened seems to remain in experience in spite of the fact that it is known to be no longer present. In this latter sense, which is an example of sensational memory, we do experience past things in memory.

The last question he addresses (4) "How do we come to know that our experience of things is not the only experience in the entire universe?" is actually the question of how knowledge can transcend personal experience. Russell's approach to the question is to focus upon not the whole of any given individual's experience, but only the experience of a given moment or point in time.

> . . . we can never point to an object and say: "This lies outside my present experience" . . hence it might be inferred that we cannot know that there are particular things which lie outside present experience. To suppose that we can know this, it might be said, is to suppose that we can know what we do not know.[23]

This can easily be refuted, as Russell does refute it with both empirical and abstract examples. We might try to recall a person's name and be certain that the name was part of our experience in the past. But in spite of our best efforts to recall it, the name is no longer part of our experience. In abstract matters, we may know that there are 144 entries in the multiplication table without remembering them all individually. The point is that there is knowledge of things which we are not now experiencing. Examples from mathematics show infinite numbers of facts and things *which do not form part of our total experience and never will.*

Thus, the scope of the domain of experience according to Russell extends beyond "my present experience." It certainly extends beyond my sensory experience, contrary to recent definitions given in the literature on consciousness.[24] Moreover, there are in the world facts *which we do not experience,* and there are particulars which we do not experience. "My present experience" consists of only *some* of the things of the world [but not all] which are collected together in a group at any given moment of my conscious life into a group. This group consists of things which exist now, things that existed in the past, and abstract facts. It is also the case that in my experiencing of a thing something more than just that mere thing is involved, and whatever that something more is may be experienced in memory. For Russell, a total group of "my experiences" throughout time may be defined by means of memory, but this group does not contain all abstract facts, and does not contain all existent particulars. It also does not contain the experiencing which we believe to be associated with other people.

2.2.2. Indexicality: A Way to Publicly Access Immediate Awareness

The differences between the two kinds of knowledge based on the distinctions we have explored thus far may be highlighted with Russell's use of the Memorial Hall example:[25]

> And when I actually see Memorial Hall, even if I do not know that that is its name . . I must be said to know it in some sense more fundamental than any which can be constituted by the belief in true propositions describing it.

Russell's concern with this example is to reject belief as a necessary condition of knowledge by acquaintance.[26] Knowledge by acquaintance is also non-linguistic and non-propositional in that the use of words asserting a proposition in a declarative sentence cannot communicate to another person the meaning and the knowledge by acquaintance of an object. That object is a *particular*, while the meanings of most natural or artificial language words, insofar as they are common to two people, are almost all *universals.*[27] Since universals are necessary for classification, and classifications are asserted in descriptively functioning declarative sentences, it follows that on Russell's account, the nature of knowledge by acquaintance (where the object known is a particular), is clearly non-classificatory.

However, certain indexicals can be used to disclose to another the object of one's knowledge by acquaintance. These indexicals reflect *speaker* meaning of those words, not word meaning. The *non-classificatory* nature of knowledge by acquaintance, where certain indexicals are used to ostensively "point" to an object of acquaintance is brought out in Russell's analysis, in addition to the irreducible nature of knowledge by acquaintance. Knowledge by acquaintance cannot be "captured" by or reduced to knowledge by description:

> If I say "this", pointing to some visible object, what another man sees is not exactly the same as what I see. . .Thus if he takes the word as designating the object which he sees, it has not the same meaning to him as to me. . .The words . .*will omit what is particular to it*, and convey only what is universal.[28]

The meaning of indexicals or language used demonstratively or ostensively, such as 'this' and 'that', is not equivalent to, nor identical with, language used descriptively. Meanings of such indexicals reflect *speaker* meanings, (actually, I would say *person* meaning, since we are not, strictly speaking, referring to *speakers* of a language) not just *word* meanings, whereas language used descriptively may be almost entirely confined to word meaning.[29]

The issue that Russell recognized here and tried to deal with is still with us. Among other things, contemporary efforts in artificial intelligence, AI, and artificial life, AL, to achieve machine translations of natural language, are stymied by context-dependent indexicals such as these. This is so precisely because those making such efforts conflate *word* meaning and mathematical functions, ignoring *speaker* meaning. But epistemological analysis shows that such indexicals disclose *speaker* meaning, primitive relations or structures in contents of thought. As such, there are powerful arguments showing that they cannot literally be mathematical functions.[30] I will later present arguments for this.

At best, Russell's knowledge by acquaintance can be communicated ostensively. That is, it can perhaps to some degree be disclosed by linguistic verbal "pointings" with indexical words such as 'this' and 'that'. If Russell had recognized knowing how and its relation to acquaintance, he might also have acknowledged that it could be communicated by nonlinguistic indexicals such as physical gestures, nods, finger pointing, and physical somatosensory-motor patterns of performance, including the manner by which one does something.

On the other hand, as Russell draws the distinction, knowledge by description is classificatory, and proceeds largely by means of class logic. In the case of knowledge of universals, the object of one's knowing can be communicated by [public language] description to another, though the something which makes one's thought of the universal a *particular* dated event cannot be communicated. If I think 3+3=6, then the object of my thought, the universal, 3+3=6, can be communicated. However, there is something that makes my thinking it a *particular* dated event, and this something, Russell says, is publicly incommunicable to another person.

2.2.3. Experiencing and its Objects

Experiencing is the most comprehensive of all the things which happen in the mental world. Conceptually, judging, feeling, desiring, and willing all *presuppose* experiencing but are different from it. Regarding the extension of present experience in time, he states that some objects undoubtedly fall "within my present experience." There are other objects that were within my experience at earlier times which I can still remember. For example, I can now be immediately aware of the setting sun's rays as they are presently configured by cloud formations on the western hillside outside my San Antonio home. This unique configuration of particulars did not fall within my experience at earlier times. However, I can also experience remembered objects. Tomorrow, I will remember the setting sun's rays I see right now.

Abstract Objects

Moreover, included within our experience, we can also think of or experience abstract facts of logic and mathematics, and we can also experience our own experiencing. Right now, in reading Penrose's wonderfully concise descriptions of Gödel's undecidability and incompleteness proofs,[31] I can experience those very abstract facts of logic and mathematics. Additionally, my experiencing of those facts as I now contemplate them, becomes yet another object of my experience.

From these facts of our experiencing, however, we can again ask how we can know that our experience is not all there really is in the world. This is the problem of how our knowledge (knowing) can transcend our own personal experience.

Russell considers *infinite* arithmetical facts as a way of showing this. We cannot think of more than a *finite* number of them during our lives. For example, the number of functions of a real variable is infinitely greater than the number of moments of time. Hence even if we could live forever and spend eternity thinking of a new function every instant, there would still be an infinite number of functions which we cannot have thought of. There would be an infinite number of facts (functions of a real variable) which cannot enter our experience. From this, it follows that what is experienced at any moment is not the sum total of the things in the world.

But his discussion here has another relevance he does not mention at this point. For example, given the influence of nominalism, we may want to consider what relation [proper] *names* and *descriptions* have to those infinite functions which "we cannot have thought" and which "cannot enter our experience." We can ask ourselves questions of the following kind: What is the smallest natural number that cannot be described to a person in words? If we assume there are numbers that cannot be described to a person in a lifetime, and if we assume there is a *least* such number, call it u_0, it appears that we may have just described a particular natural number called u_0. But u_0 is supposed to be the first number that cannot be described in words. That is, we're left with an apparent paradox, as well as an apparent confusion between *naming* and *describing*.

The Berry Paradox[32] points to problems related to the cognition and existence of abstract concepts and objects which cannot be named or formalized. This is a problem pursued in some depth by Penrose.[33] Though my concerns are largely directed to perceiving and sensing with a focus upon the sensory and somatosensory-motor systems, it may be worth our while to pursue this issue a little so as to clarify Russell's distinctions between *naming* and *describing*. It may also be helpful to become clearer with respect to how the indexical function of primitive proper names permits others access to the particular objects of immediate awareness of a subject.

2.3. Acquaintance with Mathematical Objects: Problems with Unnameables, Nameability and the Berry Paradox

The issue is with those objects which, as Russell says, "we cannot have thought" and in some sense "cannot enter our [present] experience." In part, we are still concerned with his delimitation of the scope of experience and the objects falling within it, but also with the limits of naming or labeling things, and with indexing, or "pointing" to an object of thought with nonlinguistic indexicals such as images. Again, we will address the Berry Paradox mentioned above.

For realists such as Russell, mathematical objects are abstract in the sense that they are non-temporal and hence not given by means of the senses and they are not given in the past. Though each of us may come to know certain of these objects in our past, and the objects themselves may have been discovered at a given time in human history, and our understanding of them may grow over time, the objects themselves are atemporal and independent of history. Thus the relations between a subject and these objects will not include the relations of sensation or the relation of memory, where this refers to objects having a temporal relation to a subject. However, the relation may very well (and probably does) include imagining as well as conceptualizing, and possibly other relations we may not know about. Though I do not wish to introduce the primitive relation of recognition here (since Russell does not), we sometimes speak of recognizing abstract objects or concepts not given in our prior cognitive experience, though our knowledge of other abstract objects and concepts may enable us to recognize those of which we have had no prior experience. For example, some creative mathematicians among us, such as Andrew Wiles, may use known mathematical theorems in a demonstration to establish a proof which is not yet known. At some moment in time as he proceeds, he must be able to recognize the new proof, previously unknown to all, he has just demonstrated. Indeed, this is precisely what Professor Wiles recognized when he proved Fermat's famous last theorem.

The Berry Paradox is the paradox of naming arithmetical facts which we mentioned above when considering Russell's example of infinite arithmetical facts "which do not form part of our total experience." Nonetheless, though there are arithmetical facts which do not form part of our total experience, it is suggested we can name certain of those facts. In some sense, we can name the smallest natural number that cannot be described to a person in words in the space of a lifetime. If we think of all the natural numbers that can be described by human beings, beyond a certain point there is an entire realm of natural numbers that cannot be referred to by any description short enough to be humanly comprehensible. That is, *there is a realm of unnameable natural numbers*. As explained by Rucker, a paradox results when we proceed in the following way:

> Assume there are indeed numbers that cannot be described to a person in words in the space of a lifetime; and assume that there is indeed a least such number, which we may as well call u_0. Now, it looks as if I have just described a particular natural number called

> u_0. But u_0 is supposed to be the first number that cannot be described in words.[34]

This is, he says, a version of the problem of *how we can talk about things we cannot talk about,* a problem Russell at one point dismissed as self-contradictory and chastised Wittgenstein for considering.[35] But there are a number of subtle and not so subtle facets of the above purported paradox. The first is the need to distinguish between *naming* the least natural number which cannot be described in words in a lifetime, and *describing* that number. Rucker seems to conflate the two notions as is often done by those who take proper names as disguised definite descriptions or "mere labels."[36] In the theory of no-sense proper names held at one point in time by Russell, a proper name, say u_0, simply stands for the object it names. A proper name has no sense or meaning (except *speaker* meaning) other than its standing for the object it names. In this sense, and in the above example, 'u_0' might be seen as an indexical *pointing* or referring to the object it names. It does not *describe* that object; it asserts no propositions about it.

But it may be argued that what we apparently have in the Berry Paradox is a definition by a denoting phrase. That is, 'u_0' means "the smallest natural number that cannot be described to a person in words in a lifetime." Thus, 'u_0' is defined as meaning the same as that denoting phrase. Where there is one and only one such number, x, which is the smallest natural number that cannot be described to a person in words in a lifetime, then x can be substituted in any proposition containing u_0 (without altering the truth or falsehood of the proposition). The logically perspicuous definition is 'Any proposition containing u_0 is to mean the proposition which results from substituting for 'u_0' "the number which is the smallest natural number that cannot be described in words in a lifetime."'

Moreover, we must ask whether or not the number x, which is u_0, is something with which we are directly acquainted (in Russell's sense). Russell would undoubtedly say we are not, since it is one of those infinite arithmetical facts which do not form "part of our total experience," and of which "we cannot have thought." Though one might raise objections to this claim, Russell would hold that we are not *acquainted* with such objects. If so, *then we also cannot give them proper names, not even* 'u_0'. For Russell, we can only name that with which we are acquainted. But even if we take names as disguised definite descriptions or as "mere labels,"[37] we still cannot name the number *x* which is u_0.

This will be made obvious by becoming a bit more precise about what we mean by "the smallest natural number that cannot be described in words to a person in a lifetime." We could clarify the meaning of this, as Rucker does, by stating that it must be described in a billion words. As he notes, one billion appears to be a generous estimate of the number of words a person might absorb in a lifetime. No matter what method we use, such as exponential notation, there are numbers that are really not nameable for human beings. For example, a *googol* is written as 10^{100} and a *googolplex* is defined as 10^{googol}. A *googolplex* is not nameable in less than a billion words. As Rucker points out,[38] a number *googol* digits long would easily fill space out to the most distant visible star.

He considers an alternative way of trying to determine if there is any limit to the numbers that could be described on the basis of M, multiplication, and nested iterations of length M. He considers what is called the Ackermann generalized exponential G(n,k,j) as follows:

(i) $G(1,k,j)=j{\cdot}k$;(ii) $G(n+1,1,j)$=j;(iii) $G(n+1, k+1, j)=G(nG(n+1, k, j), j)$

But G(2, k, j) is gotten by multiplying k many j's (that is j^k) and G (3, k, j) is gotten by exponentiating a stack of k many j's (${}^{k}j$ or j tetrated to the k), and G (4 k, j) is gotten by tetrating a stack of k many j's (j pentated to the k), and so on. That is, G (M, M, M) is going to be a *very* large natural number. He also considers that by nesting the definitions more than two times, one could get more rapidly growing functions and names for larger numbers, estimating that the limit of what can be done might be a number P that is greater than any H(M, . . .M), where H is a function of M arguments defined by M nestings. "The idea would be that one cannot systematically reach beyond P without using a systematic procedure that in some dimension is bigger than M."[39]

But Rucker is using the term 'name' in a sense different from Russell's use of 'proper name'. Names such as 'the *googolth* prime number' or 'the least even number greater than 2 that is not the sum of two primes', or 'the first n such that there is a string of 20 sevens that ends at the *nth* place of the decimal expansion of π', are *constructive* names in the sense that it is not known whether any of them actually names a number.

The Berry Paradox, on the other hand, states that u_0 is supposed to be the first number that cannot be described in under one billion words. It is supposed to *actually* name a number. We could require, as Rucker does, that names for numbers be interpreted in one and only one definite way, thus ruling out easy ways out of the paradox. For example, someone could claim that if 'u_0' is the name of n, then it must be the name of some $m > n$ as well, so that 'u_0' itself is actually the name of infinitely many different numbers. This would be so because each time someone starts out saying 'u_0', one could then say "but that wasn't the *real* u_0. . .what I am thinking of now is the real u_0." One would then get a bigger number, and then repeat the claim all over again. We are left with one way out of the paradox which basically states that there is no way to explicate in under one billion words what we mean by "nameable in under one billion words." Rucker explains,[40]

> Where exactly does the difficulty lie? . . .The problem is. . there is no way to describe in (under a billion) words a general procedure that will translate any string of (under a billion) words into the number, if any, named by that string of words. . there is no way for a person to describe exhaustively how he goes about transforming words into thoughts.

Or, I would say, there is no way for a person to exhaustively describe how he or she goes about transforming the objects of immediate awareness [such as certain kinds of infinite arithmetical facts] *into words*. Ultimately, what we are left with to resolve this and other paradoxes involving names, is the possibility that names such as 'u_0' are really not names [in other than the sense of an indexically functioning 'proper name' that Russell spoke of] and that the concept 'nameability' [as used in examples of this kind] is itself not nameable. The symbols we use to refer to that concept, 'n', 'a', 'm', 'e','a', 'b', 'l', 'e' *point* to the concept but they do not really reach it, as suggested by Rucker.[41]

This issue includes several other problems and other issues involving particulars, concepts, and attempts at ontological reduction which need not directly concern us here. For now, I will accept the distinction as made by Russell between proper names and descriptions, though I believe it is an open question whether or not one is acquainted with numbers such as u_0 above, and entire sets of such objects. For example, we can characterize one such set as "every even number greater than two is the sum of two primes"---even if they are in some sense beyond (some part of) our experience. Russell *seems* to have been wrong in one sense to claim that they are objects of which "we couldn't have thought" *since we nonetheless do have concepts of them.*

Mental Facts, Physical Facts

A fact is mental if it contains either acquaintance or some relation presupposing acquaintance as a constituent. Any instance of acquaintance is *mental* though the object of the relation may not be mental. A fact is *physical* when some particular, but no relation presupposing acquaintance, is a constituent of it. According to Russell, true proper names can be conferred only on objects with which one can be acquainted. On the other hand, the indexical 'this' is always a proper name because it applies directly to one object and does not in any way describe the object to which it applies. 'This' is the name of the object attended to at the moment by the person using the word. It is important to stress that 'this' object does not mean "the object to which I am now attending." 'This' is given but is not defined *by the property* of being given. Only upon reflection is it "that which is given." Moreover, 'this' applies to different objects on different occasions.[42]

The datum when we are aware of experiencing an object O is the fact "something is acquainted with O."[43] The subject who knows by acquaintance is an "apparent variable" because the subject is not given in acquaintance. Subjects are known merely as referents for the relation of acquaintance and other relations such as judging, desiring, which imply acquaintance.

2.4. The Primitive Relations

Russell limited his theory of knowledge by acquaintance to the primitive epistemic relations of *attention, sensation (all of the senses, seeing, hearing, tasting, feeling, smelling), memory, imagining*, and *conception,*[44] while also sometimes including *introspection.* To some degree anticipating Ryle's later arguments on the nonreducibility of *knowing how* to *knowledge that* he also argued that these primitive relations of knowledge by acquaintance are not reducible to knowledge by description.

2.4.1. The Primitive Relation of *Attention*

To say that 'this' is the name of the object attended to at the moment by the person using the word, points to the primitive relation of *attention.* The relation of attention is not equivalent to nor identical with the relation of acquaintance, in part due to the fact that a subject can only attend to one object (or a small number of objects) at a time. As noted above, the relation of *attention* is the primitive selecting out of an object from all other objects with which one is acquainted. That selecting is done with primitive proper names, and does not imply a reflection *about* the objects of acquaintance, for example that they have a relation to the one selecting. One is merely selecting this or that among objects with which one is acquainted. This cannot be *classificatory* because the selection does not entail reflection about the object and does not depend upon invariant properties or attributes of the object(s) selected.

Russell's concept of *attention* is emphatically not the same concept James articulated in his *The Principles of Psychology* [1890], nor is it the same concept most recently included in Francis Crick's *The Astonishing Hypothesis* [1994]. Russell's concept of attention is a *primitive* relation of immediate awareness, not a propositional language relation. Both James and Crick hold that there is no immediate relation with objects. Crick has conflated two very different concepts of attention, primitive selection and attending to something (as in *paying attention*). Also note that reflection about an object, including one's self, is not a necessary condition to awareness for Russell. This is in contrast to the position of some contemporary writers on the subject of consciousness, such as Dennett, *Consciousness Explained* [1991].

The primitive relation of *attention* for Russell is the epistemic principle of selection which underlies all indexicality, reflected in the use of such natural language indexical terms as 'this', 'that', 'I,' 'now', which are also not defined but given. That is they are primitives, and also in certain patterns of action, gestures, and the use of images which we have yet to discuss. The objects selected are, as Russell calls them, "emphatic particulars," selected out for attention. All knowledge of particulars comes from this primitive relation of *attention*.

However, universals or abstract objects such as logical and mathematical objects are also objects of the primitive relation of attention. There is some ambiguity in Russell's theory on this, but it is clear that universals are objects of this primitive relation. In fact, he discusses degrees of difficulty in the nature of attention as the object grows progressively more abstract. Like Gödel,[45] he draws analogies to and relation between attention to abstract objects and attention to concrete objects. ". . .perhaps attention to an abstract object is only psychologically possible in combination with attention to other more concrete objects, the number of which tends to increase as the abstract object grows more abstract."[46]

I will later present arguments for this, but I should also again point out that AI and AL, including neural network, pragmatist, and neo-pragmatic theories, cannot account for the use of primitive or language indexicals, what Russell referred to as "emphatic particulars." This is so in part because those theories adhere to what Russell refers to as *neutral monism*, the assumption of the *cognitive neutrality* of sensation,[47] and AI confuses grammatical with mathematical functions. We will address this issue in later chapters.

2.4.2. The Primitive Relations of *Sensation* and *Imagination*

All direct two-term relations of a subject to objects, in so far as these relations can be directly experienced by a subject, imply acquaintance with those objects. 'Sensation', 'imagination', 'conception', 'immediate memory' are words which denote two-term relations and Russell seeks to establish whether the difference between sensation and imagination is a difference in the object or in the relation. He holds that differences in the object do not concern the analytical portion of theory of knowledge. It appears that one of Russell's reasons for holding this is to clear a confusion regarding the reality or unreality, existence or nonexistence of particulars. This confusion has led to the positing of "unreal" objects, such as circled squares, and problems with "non-naming names." According to Russell, the entire conception of the reality or unreality, existence or non-existence of particulars is the result of a logical confusion between names and descriptions. Because imagined sounds or colors visualized in the imagination can be given primitive proper names, they must be on the same level regarding reality as sounds and colors seen or heard in sensation. Imagined sounds or colors visualized in the imagination are equally particulars enumerated in an inventory of the universe.[48]

However, there is nonetheless a difference between sense data and imagination data. Sense data are obviously relevant to physics since they are part of the material world. The question is how are sense and imagination to be distinguished. It is not sufficient to say that "real" objects give us data of sense, and imagination data are about "unreal" objects. Psychology is no help at all since it does not define either ‘sensation’ or ‘imagination’ as relations, and furthermore requires a knowledge of physiology to discuss these experiences. Russell maintains the two as different relations between subjects and objects, again to specifically provide an epistemological analysis of and focus on the immediate [present] experience of sensation and imagination between the subject and object.

He begins with a consideration of meanings of 'sensation' and 'imagination', again ruling out psychological approaches to understanding the differences between them. He considers James'[49] view of the function of sensation, which comes closer to his own. James states: "It's [sensation's] function is that of mere acquaintance with a fact. Perception's function, on the other hand, is knowledge about a fact." Also, from the same source:

> As we can only think or talk about the relations of objects with which we have acquaintance already, we are forced to postulate a function in our thought whereby we first become aware of the *bare immediate natures* by which our several objects are distinguished. This function is sensation.

Of course, James' use of the term 'acquaintance' differs from Russell's own use of the same word. Moreover, Russell points out that though sensation has the characteristic James attributes to it, other experiences have the same characteristic. That is, we are in fact acquainted with objects "logically similar to those of sensation in imagination and immediate memory, and with objects of another kind in conception and abstract thought." Thus we must find some further characteristic to distinguish sensation from other kinds of acquaintance, for example *conception*. An obvious difference between sensation and conception and abstract thought is that the objects of sensation are particulars. We defined particulars above as entities which can only be the subject of a predicate or as one of the terms of a relation. A particular cannot be a predicate or a relation itself. Thus, sensations are always cases of acquaintance with particulars.

However, this is still insufficient as a working definition because it is too broad. As it stands, it includes both imagination and immediate memory which are themselves relations requiring definition [or at least elucidation by means of defined terms]. There are special problems with an analysis of memory and with determining the extent to which it is included under acquaintance which I do not wish to pursue here. Again, his concept of memory was solely "mind-centered" with no concept of physical memory created from repeated activation or excitation of synapses of the nervous system [*e.g.* the Hebbian concept].[50] Nonetheless, he distinguishes memory from imagination and sensation by the fact that the object of the relation of memory is given in the past. That temporal relation with an object given earlier is missing in the relations of sensation and imagination. Thus, he defines 'sensation' and 'imagination' with the use of the primitive term 'acquaintance' as "acquaintance with particulars not given as earlier than the subject."

Not all sensation and imagination particulars have some temporal relation with the subject. A denial of this assumption of temporal relation, Russell says, gives us an intrinsic difference between sensation and imagination: in sensation, the object is given as "now" [recalling that *now* is a primitive indexically functioning proper name], as simultaneous with the subject. In imagination, the object is given without any temporal relation with the subject. No temporal relation is implied by the mere fact that imagining occurs. For example, if we call mathematical objects to mind, such as triangularity with the help of images, a particular image of a triangle, none of these objects are in time at all. The object which we imagine may even undergo processes of change, for example when we imagine a song or a poem being recited. But this does not imply that the object imagined must be contemporaneous with the imagining subject. When we recite the multiplication table, the objects are not in time at all.[51]

2.5. The Concept of *Image*

His concept of image is not equivalent to or identical with a past sense datum. However, according to Russell, when we use images as an aid in remembering, we make the judgment that the images have a certain sort of resemblance to certain past sense-data. This enables us to have knowledge by description concerning those sense data. We are acquainted with the corresponding images together with a knowledge of the correspondence. Images of past sensible objects are not themselves *in* the past, that is we cannot assign a date in time to them. For Russell, this accounts for what is called the "unreality" of things imagined. Unreality consists in their absence of a date in time.

We should diverge somewhat from Russell's knowledge by acquaintance to discuss his concept of image and the indexical uses of images. This discussion verges on enormous problems and very difficult questions which are not pursued by him, and still continue to get short shrift from researchers. When he mentions the use of images as an aid in remembering, he does not mention their use as primitive indexicals. But that is clearly how they are being used even in his own examples of attempting to recall his breakfast.[52] Moreover, in the example he gives he makes clear that the images "have a resemblance, of a certain sort, to certain past sense data," and that the images enable us "to have knowledge by description of those sense data."

That is, Russell takes it that images are used as aids in classification, as sort of "copies" of past sense data. But clearly this is not always so. Sometimes images are used as an aid in remembering, as indices of the memory, but still have *no* resemblance to past sense data at all.

The most striking record of the use of images which have no resemblance to past sense data, is an account of the mind of a mnemonist given by A.R. Luria.[53] The mnemonist is someone who had what is called *synaesthesia*, the “mixing” of the senses so that words and sounds might also have shape and colour. This made each instance of his memory distinct and unique, but also made it difficult for him to form abstractions. He would also blur unimportant differences. The point in the examples that follow is that the Mnemonist's *knowing the unique* is through indexing but not through classifying. Luria made clear that the Mnemonist was "quite inept at logical organization," thus he performed poorly if at all in acts of classifying or generalizing. Bruner, who wrote the introduction to Luria's book, indicated ". . it is a memory that is peculiarly lacking in one important feature: the capacity to convert encounters with the particular into instances of the general. . ."

The Mnemonist would form color images of human voices. He once said:

> I frequently have trouble recognizing someone's voice over the phone, and it isn't merely because of a bad connection. It's because the person happens to be someone whose voice changes twenty to thirty times in a course of a day. Other people don't notice this, but I do. . .[54]

The voice heard by the Mnemonist would "crack" into a many different colors, hence he couldn't recognize whose voice it was. As Maccia notes,[55] "this person's astounding memory resulted from his ability to recall his experience of events in complete detail. The interest here, however, is not with his feats of remembering, *but how he used the images he saw*. In order to recall lists of numbers, words or symbols, he used a device of bringing to memory a familiar street in his home town, and then taking an imaginary walk along it. He would place in proper sequence the items to be remembered on trees, gates or fences, or any convenient projection or cranny. Occasionally when he was in error, he would repeat his walk and find:

> Sometimes I put a word in a dark place and have trouble seeing it as I go by. Take the word box, for example. I'd put it in a niche in a gate. Since it was dark there I couldn't see it. . sometimes if there is a noise, or another person's voice suddenly intrudes, I see blurs which block off my images.[56]

The Mnemonist is *indexing* with images, using them as aids in remembering, but the images have *no resemblance* to what is remembered. He was not classifying. Moreover, the images used were constituted of a primary with sometimes a secondary property or feature. That is, they were colors or shapes, and sometimes images made up of both. We may think of the word 'box' as formed of block letters, spelling out the word 'box', and it is easy to think of the Mnemonist putting the block letters into a niche in a gate. But a voice can become a "blur", a fuzzy *colored solid*, which blocks off his images of surrounding items such as the word 'box' which he put into a niche in a gate.

Of course, the case of the Mnemonist may be put aside as a pathological instance, though the incidence of synaesthesia throughout human populations appears to be greater than earlier thought, when Luria wrote his book.[57] Nonetheless, the human use of images as indexicals rather than as classifiers, as well as a means of thinking, is well documented.[58] But, aside from work by Kosslyn,[59] it continues to be largely neglected in behavioral and neural research studies as well as theories of natural intelligence and theory of knowledge.

Though Russell apparently held that images were used only as copies of past experience as aids in remembering and to classify, Hadamard documented testimony from Albert Einstein about the use of images in thought. Einstein stated the following:

> The words or the language, as they are written or spoken, do not seem to play any role in my mechanism of thought. The psychical entities which seem to serve as elements in thought are certain signs and more or less clear images which can be "voluntarily" reproduced and combined.[60]

Hadamard also documented differences between those who are termed "typographic visual types," that is those persons who mentally see ideas in the form of corresponding printed words, and those who think in images, the latter including some mathematicians and logicians. The cases of Poincarè, Turing, Kekule, Einstein, and others, including Penrose,[61] who have testified to thinking in images are well known. Hadamard asks the rhetorical question:

> How can we wonder that people have been burned alive on account of differences in theological opinions, when we see that a

> first-rate man like Max Müller, *apropos* of a harmless question of psychology, uses scornful words toward his old master Lotze, for having written that the logical meaning of a given proposition is in itself independent of the form in which language expresses it? [62]

There is an assumption in much current philosophy, the cognitive sciences, psychology, and artificial intelligence, that natural and artificial symbolic languages are the sole means by which intelligence expresses itself; the sole means by which mature speakers of language think, indeed the sole means by which any thinking at all is done. Clearly, if Russell's arguments and analysis given above on the nature of experience and knowledge by acquaintance are largely correct, even given the errors made, they should lead us to the conclusion that this cannot be so. Knowledge by acquaintance is not by means of language at all.

2.6. 'Imagination' and 'Sensation' Defined

Based on his analysis thus far, Russell tentatively defines 'imagination' as "acquaintance with particulars which are not given as having any temporal relation to the subject."[63] That is, my experience or immediate awareness [acquaintance] of imagining the object, the face of a person I used to know, does not include a temporal relation with me, the subject. This obviously does not mean that the object imagined, the person I used to know, does not have a temporal relation with me, but the relation of *imagining* between myself and the particular image of that person does not. He also makes clear that the concept *imagining* does not include the experience of after-images, which are part of sensation.

'Sensation' is tentatively defined as "acquaintance with particulars given as simultaneous with the subject."[64] But Russell's analysis fails to distinguish between the two relations of *sensation* and *imagination* because an image can be given as simultaneous with the subject just as the relation of *sensation* can, and an image can also be given in a relation with no time-relation with a subject. Nonetheless, as noted, there is an undeniable difference between the data of sensation and the data of imagination; we sometimes refer to the former as "real" and the latter as "unreal." Russell decided that if imagination data are given as simultaneous with the subject, their "unreality" must consist in their failure to obey the laws of correlation and change which are obeyed by sense-data and which form the empirical basis of physics. The unreality of images simply means that they are not given with any position in time.

> An imagined visual object cannot be touched . . .images change in ways which are wholly contrary to the laws of physics; the laws of their changes seem, in fact, to be psychological rather than physical, involving reference to such matters as the subject's thoughts and desires.[65]

In summary, Russell's analysis concludes that imagination and sensation are different primitive relations to objects. They differ in terms of their relations rather than in the *objects* of those relations. If images have a time-relation to the subject, it has to be that of simultaneity, in which case there will be no distinguishable difference between images and sense data. The "unreality" of images he defines as consisting merely in their failure to fulfill the correlations which are fulfilled by sense data.

But Russell's static, non-dynamic and Cartesian "mind-centered" *knowledge* [as opposed to *knowing how*] focus on the nature of imagination and sensation is wholly inadequate. It is inadequate if for no other reason than it is incomplete. In his discussion of the "unreality" of images, for example, he has failed to consider that images of physical objects, whether the physical object imagined actually exists or not [in whole or in part, as it is imagined], as well as *images of kinds of space*, can be and are used as a kind of map to guide the patterns of our bodily actions. Images of kinds of space can be used to guide our moving and touching, including our use of probes of different kinds, including abstract probes with the mind.

He goes too far when he states of "unreal" images, "They cannot be employed to give knowledge of physics. They are destitute of causal efficacy."[66] Above, we saw that the Mnemonist uses such images quite [causally] in an effective way. Indeed, in the sciences generally and in applied areas such as medicine in particular, the use of "unreal" images to enable us to plan explorations and anticipate what we will find in unknown or unexplored spaces is pivotal in effective research and development efforts, including effective medical treatment.

The point here is that Russell failed to realize the part played by "unreal" images in other relations, such as *moving* and *touching,* which he did not consider. He also did not consider that bodily moving and touching are themselves kinds of primitive immediate awareness relations which imply but are not identical to acquaintance with objects. They are acquaintance relations themselves between subjects and objects, where those objects may include images as well as the physical things and spatial configurations of one's surroundings or environment. One's physical surroundings can also be an object [term] of primitive relations of immediate awareness such as the relations of *moving* and *touching,* as my earlier example of swimming showed.

2.7. Primitive Acquaintance with Relations Themselves

Thus far Russell has considered the kinds or species of relations involved in immediate awareness, acquaintance, with particulars. We should explore to some degree his treatment of acquaintance with relations themselves, since these are universals. Universals include mathematical truths and abstract forms as well as predicates or general terms. Keep in mind that he conceived the full scope of knowledge by acquaintance as broader than, and more fundamental to, all other human knowledge. Thus, the "species" of acquaintance or awareness were held by him to be part of our knowledge of other objects higher on levels of abstraction. But I do not wish to go too far with it. To do full justice to his complete theory would require a different book. For those readers who wish to skip over this section, you can freely pick up again at the summary or in later sections of the book.

Some of the discussion that follows may seem like needless "hair splitting." But Russell was particularly concerned to avoid all instances of even the most subtle fallacies that can easily result in disastrous consequences on a larger scale unless one pays great heed to the most hidden assumptions on the smallest scale. It is important to note that the cognitive primitive relations of knowledge by acquaintance formed an hierarchy according to the abstractness of the objects involved:

> If this most abstract object is a particular, we have sensation, imagination, or memory; if a universal, we have conception and

complex perception; if a logical form . . .we have understanding, belief, disbelief, doubt, and probably many other relations. . .[67]

He does not consider degrees of abstraction, but goes on to consider the nature of pure form as possibly the highest level of abstraction involving universals. Understanding pure form as the highest level of our acquaintance with universals is necessary because of its relation to self-evidence and logical truth. Most of what he says about our acquaintance with relations focuses upon logical form and what he calls "sense" or "direction" of relations. It is acquaintance with pure form, and the sense and direction of relations themselves that permit us to have knowledge by description of mathematics and logic. To discuss his arguments, we must introduce some of his logical terminology necessary to understand relations.

A 'complex' is anything analyzable which has constituents. If two things, A and B are related in any way, there is a "whole" which consists of the two things related which is a complex. "A is similar to B" is such a complex. There is a one-one correspondence between complexes and facts, where a 'fact' is what there is when a judgment is true, but not when it is false. Complexes are either atomic or molecular, and Russell uses the following criterion to distinguish between the two: "In the verbal expression of an atomic complex, only one proposition is involved, whereas a molecular complex involves several propositions, with such words as 'and' or 'or' or 'not'.[68] He explains that by "proposition" he means a phrase which is grammatically capable of expressing a judgment; or one which, so far as form goes, might express a fact, though it may fail to do so owing to falsehood. We will focus here only upon atomic complexes.

In complexes, there are two kinds of constituents, the *terms* related and the *relation* uniting them. All constituents of a complex are either particular or universal, and at least one must be a universal. In the relation "A precedes B," A and B occur differently from 'precedes'. In this relation, 'precedes' is the relating relation in that complex, the universal. Atomic complexes have only one relating relation, and may be classified according to the number of terms other than the relating relation they contain. They are *dual* complexes if they contain two terms, *triple* if they contain three, and so on. Also relations may be similarly classified: relations which can be relating in dual complexes will be called *dual* relations, *triple*, and so on. As Russell points out, many problems in philosophy and the sciences require the consideration of triple, quadruple, and so on *ad infinitum*, relations. One such problem requiring multiple relations (on multiple levels) is the nature of knowledge by acquaintance itself.

Moreover, all complexes have a form and constituents must have a "position" in this form. For example, "A precedes B" and "B precedes A" have the same form and the same constituents, but differ in respect of the position of the constituents. Russell later refers to the "sense" or "direction" of a relation as in the relation "A before B" as the position of the terms in the complex. He argues that it is obvious that we possess some kind of acquaintance out of which our knowledge of relations is derived, but that it is not obvious whether this is acquaintance with relations themselves or with other entities from which relations can be inferred.

There are problems, however, determining the nature of a complex, even given its constituents. For example, the complex "*x* is greater than *y*" is a different complex from "*y* is greater than *x*", though they purportedly have the same logical form and they do have the same constituents. They differ with respect to the position of their constituents. This leads to Russell's notion of the "sense" or direction of a relation, and to the ambiguity in his explanation of logical form and the sense of a relation which depends upon the position of its constituents. The distinction between the two is illustrated with the word 'before':

> The two propositions "A is before B" and "B is before A" contain the same constituents, and they are put together according to the same form; thus the difference is neither in the form nor in the constituents. It would seem that a relation must have essentially some "from-and-to" character, even in its most abstract form. . .[69]

The two differ relative to the position of the constituents in the form. To explain complexes occurring in a time-sequence, where two terms have the relation of sequence and (hence) two different positions, we can recognize them again in another case of sequence, noting that one term in one position is before while the other term in another position is after. He generalizes that given any relation R, there are two relations, both functions of R such that, if *x* and *y* are terms in a dual complex, *x* will have one of these relations to the complex and *y* will have the other:[70]

> Thus the sense of a relation is derived from the two different relations which the terms of a dual complex have to the complex. Sense is not in the relation alone, or in the complex alone, but in the relations of the constituents to the complex which constitute "position" in the complex.

He argues that the necessity to consider the sense of a relation cannot be explained away, however, by monistic theories of relations. These hold that relational propositions such as *xRy* can be reduced to a proposition concerning the whole of which *x* and *y* are parts.[71] On this view, the whole contains its own diversity, and the proposition '*x* is greater than *y*' does not say anything about either *x* or *y* but about the two together. Denoting the whole by '(*x,y*)', the proposition states something like "(*x,y*) contains diversity of magnitude." But the diversity (or sense) of a relation cannot be explained this way. In order to distinguish between (*x,y*) and (*y,x*), we have to go back to its parts and their relation. For (*a,b*) and (*b,a*) consist in precisely the same parts and do not differ in any way except the sense of the relation between *a* and *b*. "*a* is greater than *b*" and "*b* is greater than *a*" are propositions containing precisely the same constituents, and are precisely the same whole; their difference lies solely in the fact that greater is, in the first case, a relation of *a* to *b*, in the second, a relation of *b* to *a*.

But Russell's account of logical form does not seem to exclude the notion of sense. He says, for example, that when all of the constituents of a complex have been enumerated, there remains something which we may call the "form" of the complex. *The form is the way in which constituents are combined in the complex.* On the surface, there doesn't appear to be any good reason to think that the "sense" or direction of the constituents in the complex is not already given in its form. The difficulty which led Russell to consider and posit *sense* as distinct from *form* comes with the linguistic or symbolic representation of the relation. For example, above we saw the relation *xRy*. There is no problem when we symbolically represent '*x* is greater than *y*' and '*y* is greater than *x*', and note that they have the same linguistic syntactical form [Russell says they have the same logical form], and they have the same constituent terms. But they obviously are different complexes and the difference, he says, is in the *sense* of the relation. The sense is to be captured in the position of the constituents in order to account for their asymmetry.

But clearly, this has to be unacceptable as an explanation and as a ground for that asymmetry. Position of symbols in a symbolic representation to reflect a natural language expression is largely a linguistic convention, varying among languages, which may or may not have anything to do with "sense" of a relation in the *logically* significant sense Russell is concerned with. Representing a relation as R(x,y) does not tell us anything about a sense of the relation. The only way we know there is a sense of a relation is when we find out exactly what the relation is. Some relations have a sense, and some don't. Their having a sense seems to depend on the meaning of the relation itself, which is already apparent, that is exemplified, in the relation itself. The sense of the relation is already combined with its constituents in a complex. That is, the sense is already there in the form of the relation. Recall that the form of a relation is the way in which constituents are combined in a complex. Position of symbols, though often [not always] reflecting logical distinctions in our natural languages, seems too arbitrary as a basis for asymmetry among relations, and seems very unsatisfactory as a basis to explain a logical distinction. To paraphrase Frege, Russell seems to have confused position of a symbol with the thing symbolized.

He is closer to an analysis of sense by his reference to the need to distinguish between *referent* and *relatum* in order to capture the notion of a direction from the term which is a referent to the term which is a *relatum*. But this seems to depend on a geometric or spatial metaphor or notion in order to capture a logical distinction. Moreover, it is still hard to see why direction (sense) isn't *already* a part of logical form, as he has defined it. The problem may in part rest with conflating linguistic syntactical form with logical form. Alternatively, it may be the case that sense of a relation is known by acquaintance in a different sense of 'acquaintance' than logical form, where 'logical form' is identified with syntactical form. A sense of a relation, or of a concatenation of relations of certain objects may be exemplified by the relation or concatenation of relations present and capable of being used as a proper name, while logical form may not be.

Moreover, the concept of a relation is supposed to be primitive, but the concept of a *sense* of a relation appears to be more primitive still, and necessary to elucidate the nature of an actual relating relation. Thus something may be amiss: either the concept of relation is not primitive but sense of relation is, or the concept of a relation is primitive but sense of a relation is not and is somehow an ancillary consideration appended to help explicate our understanding of a given actual relating relation. But Russell's analysis with asymmetry showed that the sense of a relation appears to be more basic than the relation it is a sense of. If we have a direct immediate acquaintance with an actual relating relation in a complex, then we have an even more direct acquaintance with its sense. But this seems to require degrees of acquaintance.

Whatever the case may be, he is correct in stating that we have acquaintance with something as abstract as pure form since otherwise we could not intelligently use such a word as 'relation', or intelligently think at all. "I think it may be shown that acquaintance with logical form is involved before explicit thought about logic begins, in fact as soon as we can understand a sentence."[72]

Russell's arguments with respect to our acquaintance (immediate awareness) with the form and sense of relations, as necessary to our knowledge of mathematics and logic, recalls similar (but later) arguments by Gödel:

> Evidently the "given" underlying mathematics is closely related to the abstract elements contained in our empirical ideas. It by no means follows . . . that the data of this second kind, because they cannot be associated with actions of certain things upon our sense organs, are something purely subjective, as Kant asserted.[73]

Russell's arguments actually support that acquaintance with or immediate awareness of form and relations themselves is involved long before we know anything at all. But we also must add that if this is so, we may also have a more basic acquaintance with logical sense than with logical form, and it may be that both logical form and sense of relations in both language and doing are ostensively exemplified in part with symbols as well as signs, such as patterns of action used indexically.

Both Russell and Gödel appear to have certainly been on the right track, however, given much that we now know about multiple representations of space in the posterior cortex. If mathematics is in part the science of spatial patterns, then we have within us the primitives of abstract elements, of form and sense of relations, of which they spoke.

2.8. Summary

Russell has presented us with a theory of knowledge by acquaintance, which we take to be his theory of immediate awareness. Although relying largely on subjective, introspective methods, his theory has a formal structure, beginning with the primitive relational concept ‘experience’, which he explicated by means of the related concept *awareness* and the use of indexically functioning primitive proper names. The objects of his theory of knowledge by acquaintance include facts, particulars, and universals. He argued that the scope of the domain of experience is not as broad as the scope of our knowledge. Knowledge by acquaintance is broader than knowledge by description, and we have knowledge of facts which we do not presently experience, such as mathematical facts. He narrowed his focus to our experience of particular objects which is the relation of acquaintance [immediate awareness] and argued that there are various recognizably different ways of experiencing particular objects.

Within the general relation of acquaintance, the species or kinds of relations of immediate awareness, ways of experiencing particular objects, include the following relations which are explicated or defined in terms of the primitive concept of experience:

(a) *Attention:* the primitive relation which selects what is in some sense *one* object.
(b) *Sensation*: the primitive relation which serves to define "the present time" as the time of objects of sensation.
(c) *Memory*: the primitive relation which applies only to past objects.
(d) *Imagination:* the primitive relation which gives objects without any temporal relation to the subject.

He explained that each of these is a different relation to an object. Thus even where the object may be the same, these are different possible relations to it. He has also distinguished two kinds of objects of acquaintance: (1) mental objects in which a subject is a constituent; and (2) simultaneity and succession among objects. The latter are used in our acquaintance with time, which I have chosen not to focus upon here. In general, the simultaneous [occurrence] presentation of objects and their sequence are necessary to avoid confusing these relations of time with relations involved in sensation and memory.

'Subject' is defined as any entity acquainted with something. Thus "subjects" are the *domain* of the relation *acquaintance,* immediate awareness. 'Object' is defined as an entity with which something is acquainted, and is the *converse domain* of the relation of acquaintance. An entity with which nothing is acquainted is not called an object. A fact is mental if it contains either acquaintance or some relation presupposing acquaintance, thus any instance of acquaintance is mental because it is a complex in which a subject and an object are united by the relation of acquaintance. The object, however, need not be mental. A fact is physical when some particular is a constituent of it, but does not have any relation which presupposes acquaintance as a constituent.

Entirely contrary to certain contemporary views of consciousness, Russell holds that subjects are not acquainted with themselves. The concept "I" is explained with the primitive proper name 'this' which is the primitive proper name of an object of the relation of *attention.*

Russell sorted the classes of relations as subsets of acquaintance. However, though there is an hierarchical order among the relations themselves, the order of the relations does not constitute a taxonomy because their relation to one another is not logical inclusion. For example, the most primitive or simple of the relations is *attention* which is a prior necessary condition to the others. But the other relations are not logically included in the relation of *attention.* The above "species" of relations of immediate awareness or acquaintance form a hierarchical classification as follows:

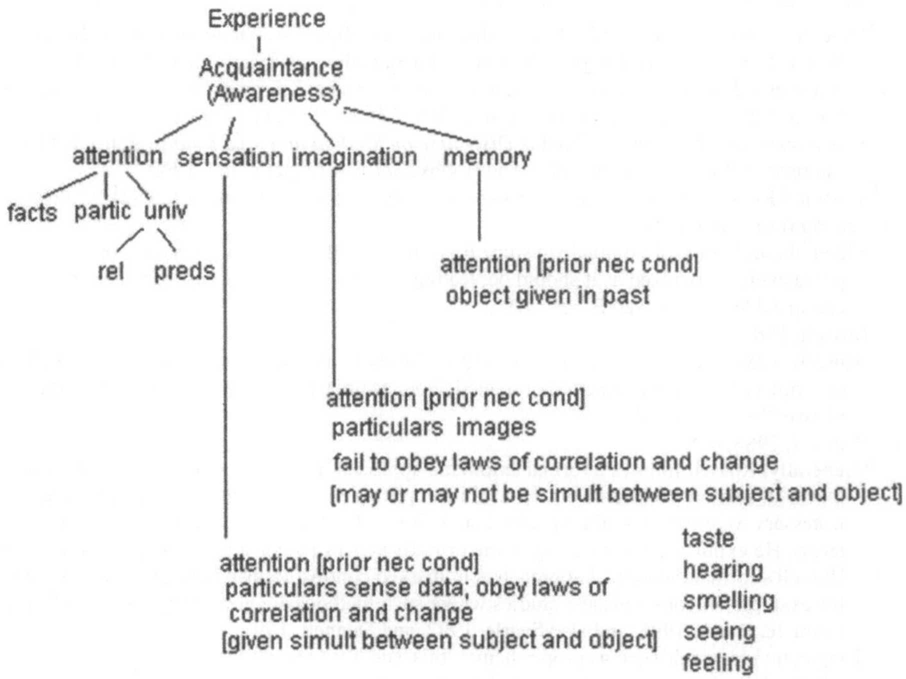

Figure TWO-1. Classification of Russell's Knowledge by Acquaintance

I will return to issues of the primitives in immediate awareness later. This will include a consideration of very complex issues involved in indexicality and individuation. We will ask very basic questions about what a thing, an ordinary object of any kind, is. We will also pursue in greater detail issues related to unique, *sui generis* objects. For now, however, we should turn to strong arguments against realism, and against our cognitive immediate awareness of anything.

---

[1] As quoted in Jacques Hadamard, *The Psychology of Invention in the Mathematical Field,* Princeton University Press, p. 142

[2] William James, *Essays in Radical Empiricism*, Cambridge, Harvard University Press, 1976 (originally published by Longmans, London, 1912).

[3] Of course, mental imagery continued to be on the margins of scientific study until very recently. The key problem was that imagery is an inherently private matter, accessible only to introspection. However, given the innovations in neuroimaging, a great deal of progress has been made in understanding the nature and role of mental imagery.

[4]Though I will deal more explicitly with this later, see Miguel A. Nicolelis, Luiz A. Baccala, Rick C.S. Lin, John K. Chapin, "Sensorimotor Encoding by Synchronous Neural Ensemble Activity at Multiple Levels of the Somatosensory System," *Science*, American Association for the Advancement of Science, Volume 268, 2 June, 1995, pp. 1353-1358.

[5]This is so even for Descartes. See his *Discourse on Method and Meditations*, translated by Laurence J. Lafleur, Indianapolis, The Bobbs-Merrill Company, Inc. 1960.

[6] Bertrand Russell, *Analysis of Mind,* London, George Allen & Unwin, Ltd., 1921, p. 110.

[7]See Penrose, 1994, p. 54.

[8] Given the influence of nominalism since Russell's day, this is a danger that is not as pervasively recognized as it should be, leading to wholesale fallacious inferences based on collapsed levels See Ned Block, 1995.

[9]Russell, 1984, p. 9.

[10]Russell, 1984, pp. 55-56. For those readers unfamiliar with the term, a predicate is usually a descriptive term that is asserted or denied about a subject. E.g. the term 'mortal' in the phrase "We are mortal."

[11]Russell, 1984, p. 81.

[12]Generally, Russell follows a formal axiomatic approach in analysis of experience and his theory of immediate awareness. He sets forth primitive or undefined terms, which are necessary to prevent circularity, and then defines other terms with the use of primitive terms. He explicates the meaning of the primitive terms by considering alternative referents. Russell's formal theoretical approach is in marked contrast to many current efforts found, for example, in consciousness studies where such methods are noticeably absent. See, for example, Block, 1995, and also Searle, 1992, and Dennett, 1991.

[13]Linguistic indexicals such as proper names 'this' and 'that' should not be interpreted as elliptical for definite descriptions. Indexicals function to "point", not to describe. Moreover, they have *speaker meaning, not word meaning*. Russell very clearly intends to distinguish between indexically functioning proper names and descriptions here.

[14] Of course, though Russell's position on these matters changed many times over the years, the most explicit treatment we have of his knowledge by acquaintance is found in the 1913 manuscript in which this is the position he took. That manuscript is the basis for this analysis. I will not assume that there is one, true position on these matters which defines Russell's final position on these matters. I do not believe there is such a position.

[15]For example, Russell's use of *proper name* as having a demonstrative (or indexical) function in his 1913 manuscript is *not* the sense of proper name that Kripke is concerned with in *Naming and Necessity*, Cambridge, Harvard University Press, 1972. [See especially his footnote #12, p. 10]. Kripke is concerned with demonstratives (or proper names which so function) which are given a reference in a definite proposition. This is also the case with Kaplan's [1989] theory of direct indexical reference. Both Kripke and Kaplan are concerned with *word* meaning not *speaker* meaning. One must keep in mind that Russell's sense of proper name in the 1913 manuscript is tied to speaker meaning.

[16] I tend to use the word 'imagining' to include 'imaging', the formation of images in the mind.

[17]There is a sense, which I will explore shortly, in which one *cannot* describe objects of immediate experience or awareness.

[18]Russell, Ibid., pp. 8-9.

[19]See, for example, Searle, 1992, especially pp. 137-138; also Block, 1995, p. 227.

[20]The assumption that recognition is *the sole* category of immediate awareness is pervasive in the Artificial Intelligence and Artificial Life communities. However, recognition can be shown to be a multi-layered concept which assumes even more primitive cognitive relations.

For my arguments against this, see Estep references, especially 1984 and 1993.

[21]Russell, p. 9, information in brackets and emphasis are mine.
[22]Russell, Ibid.
[23]Russell, p. 10. Current efforts to identify phenomenal consciousness with experience essentially take precisely this position on the nature and scope of experience. Again, see Block, 1995.
[24]Again, see Block, 1995.
[25] Russell, Bertrand, 1984, pp. 27-28 [emphasis mine].
[26] As noted earlier, knowledge by acquaintance is not the only knowing to which belief cannot be assimilated. It also cannot be assimilated to *knowing how*. See Israel Scheffler [1965].
[27] Russell, 1984, p. 29.
[28]Russell, 1984, pp. 29-30 [emphasis mine].
[29]The distinction between speaker meaning of indexicals and word meaning of descriptions is reflected in the behavioral science use of indices: an index is often correlated with an observable phenomenon that is *substituted* for a nonobservable or less observable phenomenon. We cannot observe speaker meaning, but we have access to it by means of linguistic and nonlinguistic indexicals or demonstratives. For discussion of behavioral science use of indices, see Fred Kerlinger's *Foundations of Behavioral Science*, 2nd edition, New York, Holt, Rinehart and Winston, Inc., 1973.
[30]See my "On Qualitative Logical and Epistemological Aspects of Fuzzy Set Theory and Test-Score Semantics: Indexicality and Natural Language Discourse," in *Proceedings: EUFIT '93 First European Congress on Fuzzy and Intelligent Technologies, Volume 2*, Verlag der Augustinus Buchhandlung, Aachen, Germany, 1993, pp. 585-592.
[31]See Penrose, both 1989 and 1994.
[32] Rudy Rucker, *Infinity and the Mind: The Science and Philosophy of the Infinite*, New York, Bantam Books, 1982, pp. 99-101.
[33]See Penrose, both 1989 and 1994.
[34]See Rucker, 1982, pp. 100-101.
[35] Bertrand Russell, "Language and Metaphysics," in *An Inquiry into Meaning and Truth*, London, George Allen and Unwin Ltd., 1940.
[36]See Hochberg, 1978.
[37]See Hochberg, 1978, p. 134f.
[38]Rucker, 1982, 104f.
[39]Ibid., p. 105.
[40] Ibid., p. 108.
[41] Ibid., p. 114.
[42] Russell, Ibid.
[43] Russell, 1984, p. 37.
[44] Bertrand Russell, *Theory of Knowledge: The 1913 Manuscript*, Elizabeth Ramsden Eames (ed.), Routledge, London and New York, 1984, p. 79.
[45] Kurt Gödel, "What is Cantor's Continuum Problem?" in Paul Benacerraf and Hilary Putnam, (eds), *Philosophy of Mathematics: Selected Readings*, Prentice-Hall, Inc., 1964.
[46] Russell, 1984, p. 132.
[47] M. Estep, "Critique of James' Neutral Monism: Consequences for the New Science of Consciousness," paper presented at Toward a Science of Consciousness Conference, Tucson, 1996. Abstract published in *Journal of Consciousness Studies*, Exeter, UK, Imprint Academic, 1996.
[48] Russell, 1984, p. 53.
[49] William James, *Psychology, Volumes I and II*, London, Macmillan, 1890, pp. 2-3, and also found in *Essays in Radical Empiricism*, Cambridge, Harvard University Press, 1976.

[50]See Simon Haykin, *Neural Networks, A Comprehensive Foundation*, New York, Macmillan Publishing Company, 1994, p. 49.

[51] Russell, 1984, p. 57.

[52] Ibid.

[53] Alexander R. Luria, *The Mind of the Mnemonist*, Harvard University Press, 1968.

[54] Luria, 1968.

[55] George S. Maccia,"Genetic Epistemology of Intelligent, Natural Systems," in *Systems Research*, Volume 3, 1987, p. 7.

[56] Luria, 1968.

[57] Synesthesia was first documented by Charles Darwin's cousin, Sir Francis Galton, in 1880. The condition is still poorly understood. In fact, it's not even clear how common it is, though estimates of prevalence range from 1 in 25,000 to 1 in 2000. See "Conceptualizing Through Rose Colored -Colored Senses," *Science* News Service 6 American Association for the Advancement of Science, 26 July, 2000.

[58] Jacques Hadamard, *The Psychology of Invention in the Mathematical Field*, Princeton University Press, 1945.

[59] See Stephen Kosslyn, *Image and Brain*, MIT Press, 1996.

[60] Jacques Hadamard, *The Psychology of Invention in the Mathematical Field*, Princeton University Press, 1945, pp. 142-143.

[61] See Penrose, both his 1989 and 1994.

[62] Hadamard, 1945, p. 91.

[63]Russell, 1984, p. 58.

[64] Ibid.

[65] Ibid., p. 62.

[66] Ibid., p. 60.

[67] Ibid., pp. 131-132.

[68] Ibid., p. 80.

[69] Ibid., p. 86.

[70] Ibid., p. 88.

[71]Bertrand Russell, *Principles of Mathematics*, W.W. Norton & Company, 1903, p. 225f

[72] Russell, 1984, p. 99.

[73] Kurt Gödel, "What is Cantor's Continuum Problem?" in Paul Benacerraf and Hilary Putnam, (eds), *Philosophy of Mathematics: Selected Readings*, Prentice-Hall, Inc., 1964, p. 272.

3. ARGUMENTS AGAINST IMMEDIATE AWARENESS: THE CASE OF NATURALISM

"No experiment can either justify or straighten out a confusion of thought. . ." Peter Geach[1]

Naturalist (neo-pragmatist) theories of knowledge were advanced to replace foundationalism. They were intended to replace the classical methods of *a priori* reason upon which those theories, especially Descartes' relied, and to align philosophical methods with those of natural scientists. This chapter will take a look at some of the strongest arguments for naturalism, and at some of its underlying assumptions. Among other things, we want to get clear on whether or not there is undue nominalist influence in those underlying assumptions and what the consequences of that influence might be. We also want to determine if the desire to replace what they called *a priori* reason did not also erroneously include replacing logical and conceptual analysis that is required of any reasonable inquiry.

kind of anchor, basis, or "foundation" for our knowledge of the world. Generally, foundationalists start their inquiry about human knowing with analysis of our experience in the world. Furthermore, they generally hold that our beliefs, those we hold to be true about the world, must correspond with something "out there" in the world, objectively independent of us, in order to qualify as true. And they understanding of what we hold our beliefs to be, and what we hold to be "out there," as well as in us. For about 2,500 years, those who have one way or another identified themselves as foundationalists have engaged in logical and conceptual analysis, as well as other methods, in their theorizing about knowledge and reality in general.

Though there are variations on this, most philosophers who are naturalists are what are called *methodological* naturalists. That means they generally hold that methods used in philosophy should follow or at least be consistent with those used in the natural sciences. They may differ among themselves about the extent to which philosophic methods must be like natural science methods, but generally they agree that they should not contradict one another. Methodological naturalists include those who are sometimes called substantive and cooperative naturalists.

There are also what are called replacement naturalists in philosophy, sometimes also known as "eliminative materialists." Indeed, with the introduction of naturalism into philosophy in the late 1960's, W.V.O. Quine advocated the wholesale replacement of some areas of philosophy by the natural sciences. He specifically advocated that theory of knowledge be "naturalized" and turned over to psychology.

Though the claim has been made that replacement naturalists are now few in number, their influence is nonetheless far-reaching not only in philosophy, but also the cognitive sciences and other disciplines as well. In some or all respects, most naturalists in philosophy today accept one or more of the basic principles advocated by Quine in his "Epistemology Naturalized" published in 1969.[2] Some who are very favorable toward naturalistic doctrines include Goldman,[3] Stich and Nisbett,[4] Kornblith,[5] and Harman.[6] But since Quine's work makes the strongest case for naturalism generally, though many do not go as far as he does, I will primarily focus on his arguments here.

At the outset, I should point out that I do not believe any reasonable person interested in a comprehensive, rationally informed view of human awareness and natural intelligence generally, would reject just about anything the sciences have to offer to a coherent, respectable inquiry on the subject. Of course empirical results from those sciences that address the nature of human cognition, the brain, and behavior generally, may very well be legitimately used in *any* inquiry directed to understanding human awareness, knowledge, knowing, and belief. But there are legitimate methodological differences among the disciplines. Methods of mathematicians and logicians differ from those used in neuroscience laboratories. One of the explicit goals of Quine's program was to replace what amounts to about 2,500 years of traditional philosophy with naturalist methods and goals that he set forth in 1969. Among those goals was to entirely replace methods of the sort found in Descartes. But his goals also evidently included replacing logical and conceptual analysis of the sort conducted by Russell and instituting basic principles of nominalism, undergirded by what is called his "gradualist" thesis, that are bearing destructive consequences perhaps few could have foreseen.

Russell explicitly held that there was an underlying *scientific* methodology common to both philosophy and science, and he held that scientific knowledge is central to the doing of philosophy. His methodology consisted in logical and conceptual analysis as well as making and testing hypotheses through the weighing of evidence. But with the rise of the influence of nominalism, realist approaches to issues of the kind he conducted were largely replaced by a focus upon language and language learning. The realist focus upon the relation between humans and their experience in the world was replaced by a focus upon a language interface inserted between humans and their experience in the world.

In general, realists do not deny the existence of mental states, that human beings (and probably many lower primates as well) have rich mental lives. Human beings have thoughts, ideas, desires, beliefs, and wishes in their experience in the world. Naturalist arguments in general, particularly eliminative materialists, either deny outright the existence of internal mental states or they seek to reduce those to some form of neurophysical explanation that often ends up as an issue about language.

I have found that the best way to approach the naturalist, neo-pragmatist theory of knowledge to demonstrate its stark contrast with the realist foundationalist theory, is by way of contrast between theories of *learning* and *coming to know*. Neo-pragmatist naturalists such as Quine *conflate* learning and coming to know, essentially reducing the latter to the former. This same conflation has carried over to the cognitive sciences, to artificial intelligence, and to computer systems science and engineering. However, it will be made clear in this chapter that such a reduction leads to incoherent and contradictory attempts to explain even language learning itself. Language and language learning *is* Quine's theory of knowledge, his theory of cognition, his theory of human intelligence. The focus upon language replaces the foundationalists' focus upon the relation between persons and reality.

It is crucial to understand that by 'language', Quine and other naturalists mean alphanumeric words (written or spoken) and sentences. His theory starts with sensory stimulations of receptors and observation sentences based on those stimulations. No where does Quine (or coherence theorists, as far as I can tell),[7] even begin to consider a broader concept of communication or meaning, specifically focusing upon signs or gestures that are meaningful, a kind of pre-linguistic communication system. Yet, research shows that human speech actually derives from a gestural communication system used by proto-humans and facilitated by a neural circuit found in both monkeys and humans.[8] Gestures and other kinds of non-linguistic (non-word/sentence) communication are used by humans even prior to the development of linguistic (word and sentence) systems. These are clearly signs of intelligence, signs of knowing, yet are largely ignored.

The distinction between the concept 'learning' and the concept 'coming to know' should be kept in mind as we proceed. 'Learning' is generally defined in psychology as "a change in behavior which persists and which does not result solely from physical maturation."[9] It is an operationally defined term and is central to theories of [overt] behavior. Theoretical explanations of learning focus upon processes of conditioning according to various stimulus-response models and mechanisms, the means by which behavior is "shaped." These theories have largely been based upon animal experimentation with wholesale inferences drawn from those experiments to human behavior.

Moreover, these same theories have been used to explain language learning as the shaping of appropriate verbal behavioral responses. For naturalists, learning language boils down to *copying* the verbal behavior of other speakers of the language. According to such explanations, one can learn to say the appropriate words and sentences in the appropriate order, according to appropriate rules, in the appropriate contexts, and under appropriate conditions largely by *copying* the verbal behavior of others, and being rewarded when doing so.

But characterizations of the shape or outlines of behavior do not include definitions or criteria for truth and justification. 'Coming to know', because it contains the concept 'know', must be defined in terms of standards of justification for claims to know, including standards for evidence. Where we address the meaning of 'knowledge', we must also be concerned about standards for truth. One can *learn* much that is *false*. However, one cannot be said to have come to *know* anything that is false. The two concepts are neither equivalent nor identical; hence one cannot be reduced to the other.

I will follow the same format established earlier, first explicating terms central to Quine's theory, followed by an explication. However, I will not deal with the totality of his theory, since there are portions not directly pertinent to the focus here. I will then focus upon conflicts in explanations of language acquisition, centering on the notion of learning, giving particular attention to Quine's naturalist reduction of *coming to know* to *learning.*

3.1. Definitions of Certain Terms

We begin an explication of Quine's theory of knowledge by looking to those aspects which, it will become clear, are central to his theory

Believing

For Quine, 'believing' is a disposition to respond in certain ways when the appropriate issue arises: "To believe that Hannibal crossed the Alps is to be disposed, among other things, to say 'Yes' when asked. To believe that frozen foods will thaw on the table is to be disposed, . . .to leave such foods on the table only when one wants them thawed."[10] His dispositional approach is almost a paradigm stimulus-response model used to define 'believing'.

Knowledge

To circumvent the question regarding the objects of belief, whether it is propositions or sentences which are believed, Quine redirects the inquiry to the word-pair "believes true" as relating persons directly to sentences. For Quine, knowledge is a kind of relation between persons and sentences. It is not a relationship of a certain kind between a person and an independently existing fact or other object in the world. Thus we can ask what criterion we have for saying that someone believes a sentence to be true.

He distinguishes between *causes* for believing something and evidence, making it clear that beliefs regarding evidence belong to a higher order, that is beliefs about beliefs. These guide us in our assessment of evidence. It is important to distinguish between cause and evidence since causes for believing something to be true do not establish the truth of anything. On the surface, this appears to be consistent with Russell's position. "The intensity of a belief cannot be counted on to reflect its supporting evidence any more than its causes can."[11]

Quine's epistemology is foundational in the sense that the *evidential foundation* that our whole system of beliefs must answer to consists of our own direct observations. To understand this, we should first understand what an observation is. In this, however, Quine's approach differs radically from Russell's "sense-data" approach:

Observations

> What are observations? Some philosophers have taken them to be sensory events: the occurrence of smells, feels, noises, color patches. This way lies frustration. What we ordinarily notice and *testify* to are rather the objects and events out in the world. It is to these that *our very language is geared.* [12]

Though this is not an accurate portrayal of Russell's position (an observation would clearly be an *inference* from sense data, not the sense data themselves), it is clear that for Quine, knowledge is a relation, *believes true*, between persons and language. More specifically, it is a relation between persons and certain kinds of sentences, observation sentences. For Quine, knowledge does not begin with that which is given in experience, with facts, particulars, and universals. For Quine, it begins with declarative sentences of a certain kind.

Observation Sentences

As with the object(s) of belief, Quine redirects questions regarding what an observation is to his definition of 'observation sentence'. They are sentences about external objects, expressing beliefs which do not rest on other beliefs. There are no "basic beliefs" in Quine's system. Observation sentences are those that can be part of scientific theory, affirming or refuting it, and contain terms which are intersubjectively agreeable, given an understanding of the language in which they are expressed. That is, they are sentences "whose whole occasion of affirmation, nearly enough, is the intersubjectively observable present occasion."

Terms which refer to private experience, interpretation, or meaning, are excluded. In sum, observation sentences are those "that all reasonably competent speakers of the language" will be *disposed*, if asked, "to assent to the sentences under the same stimulations of their sensory surfaces."[13] They are sentences to which we are disposed to agree to under "like" stimulations with other speakers of the language.

He begins with the observation sentence. For Quine, beliefs expressed by observation sentences do not rest on other beliefs, rather they directly "report" our sensations and are not, he claims, based on any inference. In this sense, they are also, hence, self-evident (though he does not classify them as such). There is no equivalent in Quine's theory corresponding to Russell's knowledge by acquaintance. Quine's concept of sensory stimulations is not a direct, *immediate* acquaintance with facts, constituting a kind of knowing itself. He starts with the observation *sentence* as our report *of the only evidential basis we have* for any belief, our sensory stimulations of publicly shared verbal behavior by a community of speakers of the same language.

3.2. Non-inferential Beliefs: Self-Evident Beliefs and a *Vox Populi* Theory of Knowledge

For Quine, there are two classes of beliefs that do not rest on other beliefs. As we saw above, one of these is the class of beliefs expressed by observation sentences. The other class of beliefs are those that are *self-evident.* This class consists of some logical truths, mathematical truths, limiting principles, and possibly certain moral principles. The only principles in this class which appear to rise above triviality are those of logic and mathematics. A sentence is logically true when it is an instance of a valid logical form, such as "Every A that is a B is an A." This form fits sentences such as "Every male that is an heir is a male." The logical truths are derivable by self-evident steps from self-evident truths. However, mathematical truths require the adoption of special axioms, for example the axiom of set existence, as hypotheses rather than as self-evident truths. Then one deduces consequences by self-evident inferential steps.

He also considers what he refers to as self-evident limiting principles that do not allow one or another general kind of scientific hypothesis, such as "*Ex nihilo nihil fit.*" He essentially argues against such principles because it is possible to doubt them. Of course, it is well known that the steady state theory, though in decline today, essentially repudiated this principle. Other such claimed self-evident principles include "Every event has a cause," which are also disputable. Moreover, he argues that those moral principles which are claimed to be self-evident, such as "One should not inflict needless pain," are best treated as starting points rather than as self-evident principles because it is possible to advance several different such principles which may not be consistent with one another.

We might note the contrast with Russell's knowledge by acquaintance and Quine's *vox populi* basis for his theory. On Quine's notion of self-evidence, *there is no objective correspondence between a belief and an objective fact, independent of the person who has the belief.* It is the *facts* about *meaning,* not the entities meant, that he says, that are of central concern. And facts about meaning are established in terms of *publicly approved* verbal behavior, a kind of *vox populi* basis for his theory of knowledge.

In Quine's theory, truth is replaced with a kind of popularity poll in language use. This was the first nominalist nail in the coffin of Modernism, that delicate era of the growth of human reason born of the Enlightenment. And it is a fundamental principle that has spread far and wide by those who accepted and followed Quine's "Epistemology Naturalized" since the late 1960's. His ultimate resort to *publicly approved* verbal behavior as a basis for theory of knowledge, and more broadly, a theory of human intelligence, helped ram through the pernicious and destructive logical consequences of a brand of Post-Modernism that many of us have come to sorely regret. It is this naturalistic phase of the Cognitive and Linguistic Turn in philosophy that finally twists objective truth out of the philosophical enterprise altogether.

The remainder of Quine's theory involving inferential beliefs, successful hypotheses, and analogy, are not directly pertinent to our discussion, hence I will not pursue them here.

3.2.1. A Naturalist Explanation of Coming to Know Natural Language

In general, Quine holds that all our knowledge of language is acquired by means of appropriate stimulations, conditioning the learner to make appropriate responses, to *copy*, the linguistic behavior exhibited by those who already know the language. It is a matter of changing the behavior of language speakers. It follows from this that explanations for cognitive development or growth are to be found primarily by examining the stimulatory conditions *external* to the learner in which the conditioning process takes place.

Thus far, a counter-argument to Quine's explanation might proceed along the following lines: it is not sufficient to say that learners or speakers of the language find certain linguistic expressions acceptable and not others, or that they have dispositions or predispositions to act in certain ways linguistically. We must also say *why* learners or speakers of the language have these dispositions or predispositons *and not others*. To borrow an analogy: According to the neo-pragmatist naturalist, the relationship between linguistic theory and linguistic behavior of human speakers is analogous to the relationship between a theory describing the trajectory of projectiles and the behavior of projectiles. But this is wrong. The relevant difference between human beings and projectiles is that to explain the behavior of a human being one has to explain the role played by some internal, cognitive or mental, structure.[14]

What is at issue in his theory is an explanation of how human beings *come to know* the rules of natural languages.[15] Quine claims that the rules are directly evident and hence not a matter for justification, or that they are matters of perception. In either case, learning and coming to know [natural] language are held to be a matter of conditioning directly to stimulation and have nothing to do with the internal (mental) structure of learners. The opposing argument, on the other hand, might focus on what the learner brings to the learning process, that is, what the learner already has when he or she begins to learn language, or anything at all.

Thus it is readily seen that the models of natural language acquisition explanation by these two opposing camps differ immensely. As is already evident, the two positions differ relative to whether or not there is a foundational cognitive structure in immediate awareness, knowledge by acquaintance, discussed above, or some kind of basic cognitive structure already in the person. Quine rejects such a foundation, replacing it with a naturalist stimulus-response structure.

Quine's Theory of Language Learning: *Induction* as *Learning* as *Coming to Know*

Quine's neo-pragmatist theory of knowledge, mind and meaning is succinctly stated as follows:

> With Dewey I hold that knowledge, mind, and meaning are part of the same world that they have to do with, and that they are to be studied in the same empirical spirit that animates natural science. There is no place for a prior[i] philosophy.[16]

Know*ledge*, for Quine, is made up of *linguistic* reports, verbal or written, of our sensory observations, and analytical hypotheses. The observation reports or "observation sentences" are clearly the most important, he says, since they are what we learn to understand first. The only evidence for any kind of knowing is sensory evidence. It is also the case that for Quine the process of coming to know (which is identified by him with learning) is by means of induction, which he identifies with stimulus-response conditioning:

> We learn [observation] sentences by hearing them used in the presence of appropriate stimulations publicly shared, and we are confirmed in our use of them by public approval in the presence of similar stimulations . . .The *learning process is the process of induction*. . .The generality reached by our induction is rather a habit than a law . . .What we learn by induction is a full range of scenes or stimulatory situations to which the word is appropriate---in short, its stimulus meaning.[17]

He continues by setting forth the central theses which he is concerned to show, by asserting the characteristic of language as a 'social art' "which we all acquire on the evidence solely of other people's overt behavior under publicly recognizable circumstances. ". .that, as Dewey stated, 'meaning. . .is not a psychic existence; it is primarily a property of behavior'."[18] Further, to support his contention with regard to the relativity of ontology, "It is the very facts about meaning, *not the entities meant*, that must be construed in terms of behavior."[19]

In sum, learning and coming to know words, both the phonetic (articulations) and the grammatical parts, including the meanings of the terms themselves, are relative to the sensory stimulations of particular speakers, and also relative to the sensory stimulations of the community of speakers of a given language. The way one arrives at or comes to know the *meaning* of *any* particular word, Quine explains, is through sensory stimulations, the conditioning or induction, the *copying* of the linguistic behavior of those who already know the language. One copies until one acquires, by the "process of abstraction," the habit, that is the expectation, which governs the use of the term for particular language communities .

With this explanation of word meaning, he introduces yet another problematic concept, the process of "abstraction." Of course, the process becomes more complex and difficult in cases where words do not ascribe observable traits to things. And when they do not, it is at this point he then argues for the *ultimate inscrutability of both word meaning and reference.* This is also the point where Quine's empiricism differs from more traditional empiricist positions, in that he denies any precise distinction between analytic and synthetic statements, and eliminates the traditional consideration of sources of evidence for those statements. He adopted a "gradualist" position, effectively denying that there is any other possible source of evidence other than sensory stimulations. That is, all we can claim to be certain of are our own sensations and the sentences which "directly report" on our sensations, observation sentences. And these turn only to sensations for support or justification, not to other sentences.[20] There is no place in Quine's theory for Russell's concept of knowledge of those objects and entities which exist but which form no part of our experience.

When we hypothesize about the meanings of terms in language to form generalizations to explain our use of words, such as the principles of identity and individuation, Quine wants to say the hypotheses do not resolve the indeterminacy of the meaning of the expressions. This is so because there seem bound to be systematically very different choices, all of which do justice to all dispositions to verbal behavior on the part of all concerned. He states: "When on the other hand we recognize with Dewey that 'meaning . . is primarily a *property of behavior*, we recognize there are no meanings, nor likenesses nor distinctions of meaning, beyond what are implicit in people's dispositions to overt behavior."[21]

3.2.2. Learning as a Process of Induction: A Spurious Concept

The characterization given by Quine of the learning process as a process of induction is a pragmatic notion. It is testing out the consequences of one's language acts in the community of speakers of the language. On the basis of past experience, the past language acts midst the approval or disapproval of language speaking peers, one forms hypotheses which predict future language behavior of others towards oneself as a learner of the language:

> The other fellow has affirmed or assented to the observation term or sentence in question, or has approved our assent to it, amidst various scenes that were similar to one another; and we predict that he will do likewise in similar scenes hereafter.[22]

The learning process as a process of induction also depends on an "inner sense" of "subjectively natural kinds," that is on a sense of similarity. The learner of the language must sort out language acts from other experiential events and sort out approval of his or her own language acts by others from disapproval of them. One must *also* be able to discern *superficial* from *non-superficial* degrees of similarity, though Quine doesn't appear to address himself with this problem.

> The learning process depends, like any, on a prior sense of similarity, a sense of subjectively natural kinds. We volunteer or assent to the observation sentence 'Water,' in some stimulatory situation, and expect the other fellow's approval of our progress in the language, only because this stimulatory situation seems like the one he was enjoying when he said 'water' or assented to it.[23]

What Quine requires is that the learner *single out for attention from other features of the environment* some *single* feature given in sense experience. That is, the method of induction, he claims, is the psychological process of *abstraction* from experiential events to general concepts or universals.

We should pause to carefully evaluate what he has stated so far. It is specifically these claims regarding the doctrine of abstractionism, the "lynch pin" which supports what he claims is the method of induction, the processes of learning, and his explanation of meaning, with which one must take issue.

This doctrine of abstractionism breaks down when we carefully analyze the formation of logical, arithmetical, and color concepts. The inadequacies may be demonstrated by examining his appeal to "an inner sense" to explain the use of logical terms, and in his claim that there are no special logical concepts which apply to logical words (hence holding there is no strict distinction between analytic and synthetic statements).

According to Quine, the learning process, which is the process of induction, depends upon a *prior sense of similarity of subjectively natural kinds*. That is, it depends on the "inborn propensity" to find one stimulation qualitatively more akin to a second than to a third. Thus, it is according to this inborn sense of similarity that the "stimulus meaning" of the term, in its variety of usages, is supposed to be held together.

However, this does not explain learning even the simple logical concept of negation. Geach[24] offers the following criticism:

> If 'not' in 'not red' were merely a signal for a special exercise (that is in the presence of appropriate stimuli) of the concept *red,* our grasp of its meaning in such schemata ["if every P is M, and some S is not M, then some S is not P"] would be inexplicable.

Geach's point is that the ability to use the logical concept of negation cannot be explained by an appeal to some "inner sense" or feeling of "subjectively natural kinds" because *there is no* characteristic feeling or sense of otherness associated with negation. The concept of negation cannot be learned by means of what Quine calls the inductive process.

Moreover, general color concepts can't be learned by Quine's concept of induction either. It doesn't make sense to confine use of such general terms as 'red' to situations including "stimulations" by some object to which the term applies. The term 'red' can be used in situations *not* including anything red. Additionally, the same arguments can be used against such explanations of mathematics learning. In effect, elementary mathematical concepts cannot be learned by sensory stimulations and the copying of other people's language behavior.

One primary point against such explanations is that there is more to the "presence of similar stimulations" than genetic inheritance. In order to copy or mimic anything one must first sort out *that which is similar from that which is not similar* throughout continuously changing contexts. Learning the meaning of similar stimulations *presupposes cognizance of,* not merely *discrimination of,* stimulations. *That is, such an explanation already assumes the intelligence it sought to explain.* Quine's explanation, and any eliminative materialist explanation making such prior assumptions, is guilty of begging the question.

Humans are just not passive recipients of external stimuli, as numerous studies have shown.[25] Indeed, research projects in areas of neuroscience are addressed to the mechanisms determining how the brain "decides" to attend to some stimuli (as in the preattentive phase) and ignore others. These are issues that have yet to be resolved in the sciences. Moreover, a person's emotional state may have a great deal to do with what their brain selects out for attention while ignoring everything else in the stimulus field. Some studies suggest that "our perceptual systems are exquisitely tuned to the occurrence of emotionally significant stimulus events, requiring much less attention or effort to reach conscious awareness compared to events of neutral [emotional] value. ."[26]

Additionally, as with all the naturalist explanations I have seen, Quine ignores indexicality entirely. This is due in part to his almost total focus upon declarative language statements as the sole linguistic expression of cognitive significance. Humans, like monkeys, clearly communicate with limited vocalizations, various sounds, and a very large repertoire of oral-facial gestures. In the monkey, these are linked to the "pre-Broca" area of the brain. That area in the human brain and its counterpart in the monkey have neural structures for controlling oro-laringeal, facial and hand-arm movements, all of which are used by human beings to communicate meaning.[27]

In addition to reading alphanumeric expressions of language, human beings also "read" meaning in the body behavior of others, a fact assumed but nowhere explained even in Quine's theory. Reading facial expressions, for example, has a lot to do with our success and ability to communicate with one another,[28] even when we are speaking and writing a natural language. Only those with an extremely narrow concept of language would limit cognitive meaning to symbolic (word) expressions and vocalizations having a certain form.

3.2.3. Two Concepts of Induction

Moreover, as with many other naturalists and materialists, Quine has based his concept of learning entirely on an entirely spurious sense of induction. There are *two* senses of the concept 'induction' which must be distinguished from one another. There is induction *as statistical form of argument* and there is induction *as process*. Quine adopts the latter. That is, he takes induction to be a process of reasoning in which one somehow derives "theory" or an hypothesis from [sense] data that is somehow "received" by a passive recipient.

This sense of induction can be traced back to Francis Bacon,[29] who presented induction as a way of discovering truth. For Bacon, as well as for Quine, through the supposed "process of abstraction" from particulars, generalizations about the world can arise. That is, induction as process or learning takes place. But this is an unsupportable notion of induction because the so-called process of abstraction breaks down. It breaks down because already assumes what it purportedly seeks to establish. It is yet another instance of begging the question. Induction as process is erroneously taken to be a kind of logic of discovery when in fact it is a logic of verification.

Induction is in fact a kind of statistical inference or reasoning which is involved in determining whether theory is supported by data. That is, it is a kind of *statistical* argument. Through induction, one makes an inference from *some* instances of a collection to *all* instances of that same collection. The conclusion makes a claim which goes beyond the premises, making the conclusion only probable and not logically necessary. It is induction *as statistical inference* which is the true sense of induction, not induction *as process*. When induction is taken in the sense as a statistical form of argument, it rules out other spurious senses of the kind involved in Quine's theory.

3.3. Indeterminacy of Translation and Other Problems

The above are not the only problems to be found in Quine's and other naturalists' theories. As pointed out, one of his primary theses is the inscrutability or indeterminacy of meaning and reference. Bradley[30] argues that if indeterminacy of translation and the possibility of multiple systems of interpretation are the case, as is certainly the case in Quine's theory, then relativizing the use of language to any particular reference frame *will not adequately explain how any one ever knows what they are talking about.* This is so, Bradley argues, due to the regress which develops with Quine's notion of relativizing the original linguistic corpus to reference frames (background language) and his failure to stop or render harmless the regress in light of his thesis:

> Now I seem to understand the sense--and thus the reference system--of the background language, I seem to be able to intend sentences formed in it in a certain way, and I thus seem able to fix for myself a reference system for the original linguistic corpus. *But that counts for nothing* for I seemed to have precisely the same capabilities with respect to the original linguistic corpus, yet was forced to regard them as illusory by the argument for the variety of interpretations.[31]

Bradley demonstrates that Quine's position is untenable as it relates to the thesis of intensional and extensional inscrutability of language meaning. It is obvious that for Quine there is no single sentence on which all speakers of the language would agree with regard to its meaning. With this, he pounded the second nail in the coffin of Modernism, put there by this and related brands of naturalism. As a consequence, we are indeed left *without an explanation for how anyone ever knows what they are talking about.*

There are still further problems with Quine's position. He claims that meaning and belief are to be found only in dispositions to overt behavior, in particular dispositions to overt verbal (linguistic) behavior. In addition to logical problems included in *any* appeal to *dispositions* to explain behavior,[32] we might reasonably expect to look to the conditioning or inductive process by which these dispositions are acquired to find explanations of that process. We should expect to find an explanation and hence to understand all there is to understand about how meaning and knowledge are acquired. But we do not find any explanation at all; we only find more problems.

To explain this process, as we saw above, Quine set forth his notion of the observation sentence that is supposed to directly report on sensation. The important trait of observation sentences is not that they have to actually be learned ostensively, but that they *could* have been learned ostensively. He states:

> And it is a trait that is socially traceable, for what it comes to is just that all speakers of the language, nearly enough, will assent to the sentence under the same concurrent stimulations. . .speakers of the language will ordinarily agree as to the truth or falsity of an observation sentence when they are stimulated alike.[33]

The sets of beliefs formed by direct stimulation and those beliefs removed from direct stimulation are considered "true" or adequate so long as they tend to support our expectations, that is, so long as we all get stimulated more or less alike. With Quine, truth is no longer a necessary condition to know or (justifiably) believe anything. Only "like stimulations" are required.

But suppose, as Bradley points out, that I have a set of beliefs which calls for not believing that anyone has ever been on the moon. My neighbor, an astronaut, however, holds the belief that someone has been on the moon because he was there. Suppose further that my beliefs call for also believing my neighbor is a liar and that all references to anyone ever having been on the moon are fantasies or lies. My set of beliefs *do* render my stimulations coherent, *do* support my expectations. What reason do I have to question my beliefs? Indeed, one might with a great deal of justification point out that many who call themselves "Post-Modern" are asking this question for precisely the reasons Quine gave to them.

One could say that my neighbor and I are not sharing "the same concurrent stimulations" and hence were not part of the same language community. But what exactly does it *mean* to say that a number of persons *share the same or like concurrent stimulations* and are part of the same language community? How much does this explain anyway? In addition to the faults pointed out by Bradley above with this line of Quine's theory, could the notions of 'condenser' and 'deertrack' or language meaning in general, be learned ostensively, by direct conditioning to stimulation as Quine claims?

Some facts regarding stimuli and sensations may serve to cast very serious doubt on whether we should turn to sensations or dispositions to overt behavior to explain meaning and knowing at all. As Kuhn explained decades ago,[34] people do not *see* stimuli, but *have* sensations. Moreover, there doesn't appear to be any compelling reason at all to suppose the sensations of any two people are the same. He says,

> . . .much neural processing takes place between the receipt of a stimulus and the awareness of a sensation. Among the few things that we can know about it with assurance are: *that very different stimuli can produce the same sensations*; that *the same stimulus can*

produce very different sensations; and finally, that the route from stimulus to sensation is in part conditioned by education.

Furthermore, numerous research studies in neuroscience show that we must clearly distinguish between activity at the level of sensory receptors caused by, say, a moving object, and *thought about* that object. We know, for example, that even *thinking* about a moving object can cause areas of the brain's motion detection system to activate *even before* that moving object appears to our sensory receptors.[35] This phenomenon and its meaning cannot be explained on Quine's model of language learning.

Additionally, as already noted above, determining how the brain "decides" to attend to some stimuli and ignore others is a critical issue in neuroscience because, contrary to Quine's assumption, human beings are not passive receivers of stimuli. Sensory receptors in the brain are influenced by many factors, including one's emotional state. These factors can and do directly influence what one perceives; indeed, they influence whether or not one perceives a given stimulus at all.[36]

The issues Kuhn raised, however, point to a more fundamental problem with Quine's explanation of the indeterminacy of meaning and translation. There are also problems, in addition to the ones already pointed out, with his notion of observation and observation sentence. Underlying these problems is a major recurring theme of Quine's work, his claim that all *intentional* phenomena such as meaning, belief, and desire are underdetermined by all possible evidence. We cannot determine *from all possible evidence* whether two people meant, believed, or desired the same thing. He further argues that because meaning cannot be uniquely determined by all the evidence, intentional usage should be excluded from science. This would effectively rule out the very kind of neuroscience research inquiry just referenced above, since that research dealt with intentionality and meaning.

To bolster his arguments, he holds that the notions of observation and observation sentence are "clear and clean-cut" and provide, *unlike intentional concepts,* the scientific basis for belief claims. But his notions of observation and observation sentence *also* possess traits of intentionality and indeterminacy which he would wish to reject from his theory.[37] This undercuts a substantial foundation of his own theory. Recall that it is the observation sentence that is supposed to directly report our sensory stimulations. These in turn are the evidential foundation on which our entire system of beliefs rests. Beliefs expressed by observation sentences do not rest on other beliefs. The fact that sensory stimulations are embedded with intentionality may be brought to light by looking more closely at his own characterizations of them.

3.4. Are There Immaculate Sensations?

Quine characterizes observations in terms of objects or events out in the world that we "ordinarily" notice. The things we observe are not sense data such as smells, feels, color patches, *etc.* but the things to which our *language* is geared. Observation sentences will include "terms that we can all apply to their objects on sight: terms like 'mailbox', 'stout man', 'gray moustache,'. . ."[38] They are not terms which depend on our past experience and which are shared by only a few. They have two traits: (1) They can be "checked on the spot." That is, an observation sentence is intersubjectively verifiable in that we can depend on other witnesses to agree to it at the time the object or event is described. All speakers, nearly enough, will assent to the same observation sentence under the same concurrent stimulations. (2) An observation sentence *could* have been learned in the sensible presence of something the term describes, that is by associating heard words with things simultaneously observed.

However, Quine's assumption that a community of speakers of the same natural language *has relatively the same sensations when stimulated alike* is not warranted, as Kuhn pointed out, and we noted above. It is also contradicted by a great deal of neurological research.[39] We know that humans do not give equal access for entry into consciousness to all stimuli. We do ignore some events, words, and stimuli, but not others. Moreover, as noted, we are influenced by emotional meaning of stimuli. Quine has built the edifice of his theory of knowledge upon the presumption that sensations are cognitively "neutral," immaculate or clean of any intentionality or meaning on the part of the one feeling or having the sensations. Intentionality or meaning are tied solely to language, observation sentences, in his theory. Sensations have no meaning, hence no intentionality, according to him. Moreover, for a sentence to count as observational, it must have both traits characterized above. If a sentence fails to meet either condition, then it is non-observational. Furthermore, a sentence may meet both criteria for a given community, say, a community of experts, hence it will be observational for that community; but the same sentence may fail to meet both criteria, and hence be non-observational, for another community, say a community of non-experts.

The crux of the issue whether or not observationality is intentional and hence as indeterminate as belief, desire, and meaning, hinges upon whether or not it is "clear and clean-cut" when a sentence can be learned ostensively and when there is intersubjective agreement under "the same concurrent stimulations." Martin[40] challenged the second part of this, arguing that "the same concurrent stimulations" is as indeterminate as belief, meaning, and desire.

3.5. Matching Up Stimulations

Quine's clear and clean-cut observationality is obviously not immaculate. Intentionality, meaning, belief, and emotion are involved in observation. Observationality cannot be rendered purely extensional by an appeal to stimulatory patterns, states of receptors, or forces impinging upon them.

Moreover, translation does not begin by *matching up stimulations.* Neither the knowing and observations of persons, nor meaning are reducible to an indeterminate translation of stimulatory patterns. As argued above, *a prior cognitive apparatus is involved to even begin to select or sort out one sensation or stimulation from another, let alone one object from another.* Recent research in neuroscience referenced above give us every reason to believe this is so.[41]

But this brings us back to Kuhn's argument. According to Kuhn not only is it the case that very different stimuli can produce the *same* sensations and the same stimulus can produce very *different* sensations, the route from stimulus to (awareness of) sensation is in part conditioned by education. That is, meaning is also involved. This undercuts reliance on neural processing to understand the nature of meaning, belief, other intentional phenomena, knowledge and knowing.

It definitely rules out a *reduction* of knowledge, knowing, and belief to neural processing of sensory input.

3.6. Are Meaning Structures Equivalent to Neural Structures?

Even if we assume that any given language community,[42] has relatively the same sensations when stimulated alike, could we safely assume that learning of that language can be achieved ostensively, by conditioning or induction, the process of abstraction and generalization from observed use? The answer to this is surely that we cannot. Detailed empirical evidence will be presented in a later chapter showing that we have cognitive structures (that is, meaning structures) which are already operative whenever we consciously or even *unconsciously* distinguish anything at all, including sensations. These cognitive capacities cannot be adequately explained in terms of, or reduced to, those sensations, or in terms of our physiological responses. To attempt to do so would be to assume the very thing one seeks to explain. Moreover, very different cognitive responses can result even from the same stimuli; this means that any attempt to explain cognitive capacities in terms of sensations would result inevitably in contradictions.

Quine attempted to use the notion of observationality and observation sentence to replace the foundationalist's knowledge by acquaintance or basic belief. It is crucial to see that just as knowledge by acquaintance and self-evidence form the basis on which all other knowledge and belief rests on the traditional foundationalist explanation, so for Quine, it is the observation sentence and the stimulatory situations and patterns on which everything else in Quine's theory rests. This results in begging questions by assuming some of the very things one seeks to explain. It also results in outright contradictions. Thus, he ultimately has no firm basis for his theory.

3.7. Critique of Naturalist Theory of Knowledge

Recall that for Russell, knowledge by acquaintance consists of *direct awareness* of an object. It is an *immediate--not* mediated--

relation with an object, such as an independently existing fact. For Russell, there is a variety of facts with which we are acquainted through the inner and outer senses, starting with sense data, and ending with those facts with which we are acquainted through conceiving, universals.

However, Quine conceptualizes all this differently. He replaced Russell's knowledge by acquaintance with observation sentences and stimulatory patterns. He points out that there are two levels to distinguish here: (l) the level of *input* of "unprocessed information" where, he suggests, we do well to speak not of sense data but of *nerve endings*, and (2) the level where the information has been processed. This is the level of conceptualization where he says we do well to speak of *observation sentences* and not of sense data. One might object that the concept 'unprocessed information' as input simply doesn't make any sense, since the process of inputting information *is* to process it.

Moreover, along lines I have argued above, it also does not make sense to collapse the conceptualization of a thing with having an observation sentence to assert about it. As a reminder, we should keep these levels and this model in mind when we look at the design of architecture of intelligent systems later, since some systems engineers have apparently followed Quine's approach somewhat faithfully. These AI models share some of the same problems with the theoretical bases of their design.

Recall that for Quine there is no sharp distinction between analytic (logical) and synthetic statements. Analytical hypotheses, he says, are arrived at gradually. 'Gradually' denotes the distance or remoteness of a term from the data or sensations in terms of whatever is relative to best explain our stimulations at any given time. However, this notion is based upon his spurious notion of induction, which we saw breaks down because it is based on yet another spurious process, the process of "abstraction." Moreover, as we also saw above, Quine clearly maintains that there can be different choices in word meaning "all of which do justice to all dispositions to verbal behavior on the part of all concerned." He makes the mistake of assuming that the subject matter of grammar or the theory of grammar *is* the inborn propensities of speakers to find that certain expressions sound similar to others, rather than that the subject matter is the structure of language itself. As pointed out by Graves, Katz, *et al.* [1973], this is like saying that because sensory experience constitutes evidence in physics, therefore the theories in physics are theories about sensory experience. It also recalls Russell pointing out the fallacy of assuming that because we see the physical object, the retina, we therefore know all there is to know about the physical knowledge we have derived by seeing it. These are instances of the fallacy of assuming that which one seeks to explain, the fallacy of begging the question.

In fact, it seems to be on the basis of the above assumption that Quine builds his theory of *logical truth* that includes the elimination of the strict distinction between analytic and synthetic statements, his "gradualist" thesis.[43] But there are clearly examples of statements which one would want to consider as analytic or *a priori*. Quine defines as analytic any statement which is reducible to a logical truth by interchanging synonyms,[44] and cites an examples of this, 'A bachelor is an unmarried man.'

Learning Colors

While not going too far into a discussion of degrees of analyticity or recalling the *a priori*-synthetic debate, one may ask how Quine would view learning or coming to know the meaning of the following sentences containing color concepts:

> (1) Nothing can be red and green all over at the same time for the same viewer.

or,

> (2) Nothing can be red and unred all over at the same time for the same viewer.

On Quine's theory of knowledge, the question to be asked would be whether these statements are synthetic or analytic. It is important to point out that one would *not* ask, on Quine's theory, if they are *a priori* synthetic statements. For the observer who would say they are synthetic, adopting Quine's explanation of knowledge, one would assent to the statements as true *synthetic* statements because one *had*

never encountered any object which was both red and green all over at the same time; and one *had never encountered* any object that is both red and unred all over at the same time.
Moreover, one would say that the language community of which the observer was a member had also not encountered such objects. And one might argue that such a discussion would ultimately have to hinge, in part, on the meanings of 'red' and 'green' for (1) and 'red' and 'unred' for (2). For the observer holding the statements to be synthetic, the matter is resolved by pointing to things having the colors red and green or being unred.

But one might argue that being red also *means* being *not green* and being *not unred.* This response is objectionable, on Quine's view, however, because one can know what 'red' and 'green' and 'unred' mean *without knowing they are incompatible properties.*

Arguing further, we might ask if it possible to learn colors this way. That is, (on Quine's theory) first learn the purported specific sensations to which these words apply (assuming for the sake of argument this is possible, and putting aside arguments by Kuhn above showing that it isn't), and then later to learn their incompatibility? Clearly, the problem is at least *in part* a word-semantical one. It seems that the compatibility and incompatibility of colors is integral *to the very meanings* of color concepts. If this is so, then it is integral to the learning of colors themselves.

But the situation is actually far more complex than this. As Geach[45] points out, *the learner must already be in possession of some rather sophisticated concepts to be able to make any sense out of the language learning, including color learning situation, at all.* For example, Quine's theory requires that the learner single out for his or her attention from other features of the environment, some one feature given in his or her sense experience. That is, "abstracting" that feature *from other features given simultaneously*, and forming, by virtue of this abstraction (including the observation of the linguistic behavior of other members of the language community), the appropriate concept of that feature.

But the above color statements are minimally good candidates for *a priori* synthetic statements, in part because they cannot be reduced to semantical (meaning) principles about the use of color words, nor can they be reduced to sensory stimulations and observation sentences, which we showed in earlier arguments. Quine's theory is deficient in part because it cannot account for these statements. He cannot account for learning color concepts.

Contrary to Quine's model, as argued by Kuhn, from what is known of neural processing, we cannot rely on looking to sensations and stimuli to tell us how one learns, what one means, believes, or knows. From early experiments in the 1960's,[46] to more recent neuroscience research, it has been demonstrated that discriminating features of the environment is already part of our neural makeup. It is even apparent in day-old babies *without* conditioning. Though the early research was readily available to him, Quine nonetheless opted for a stimulus-response, mechanistic, discrete, linear and additive model to explain human learning and theory of knowledge. Furthermore, he attempted to account for the learner's disposition to behave linguistically in certain ways according to grammatical rules on the spurious basis of the learner's "inborn propensity to find one stimulation qualitatively more akin to a second stimulation than to a third."

But neurological research strongly supports earlier experiments that indicate that learners bring cognitive, logical structures or capacities *to* the stimulations. They bring these structures to bear on sensations they actively *select* (in both the preattentive and attention phases) as well as receive, in order to make any sense or meaning of those sensations at all. They do this long before verbal linguistic behavior is developed and *before* any conditioning or so-called inductive process is incurred. Even newborn babies are drawn to face-like stimuli, and normally developing children as young as 6 months old show different brain activity when they see their mother and when they see a stranger. Tests show that they recognize their own mothers.[47] Earlier, Kessen [1965] provided even more evidence. He examined the manner in which day-old babies scan with eye movements a solid-colored triangle on a contrasting field. Indeed it is quite apparent from the tracings obtained that neonates are able to locate corners and lines, along with a sharp contour separating dark and light; and what is just as important, there is not only discrimination at the start but also appropriate motor activities. . .[48]

Bowers[49] also conducted early experiments on human infants from four to nine weeks of age in which objects were moved behind a screen and then made to reappear altered form. The findings suggest that the *principle of identity,* perhaps in only primitive form, is already operative this early in life. These discriminations have to be regarded as meaningful ostensive (indexical) demonstrations of attention, in Russell's sense. They must be regarded a kind of basic or immediate awareness, that is not the result of conditioning to stimuli.

3.8. Summary

Because of the widespread influence of Quine's theory, particularly on neural network theories and GOFAI in general, more critique and evaluation are necessary. We may note in favor of Quine's theory the theoretical simplicity he has introduced. The things given in experience are stimulatory sensations. There is no Cartesian dualism of mind (the mental or "ideas") and matter, hence Occam's razor would indicate preference for his theory over more complex ones. However, simplicity must always give way to truth, and where there are problems explaining the facts, as Castañeda would say, we must complicate the data.[50]

Thus, in addition to problems we raised earlier with his attempts to get rid of the intentionality of observation and observation sentences, due to the incoherence of his notion of 'same concurrent stimulations', we must also again indicate that there are additional serious problems with the assumed *neutrality* (that is, *non-intentionality) of sensation.* This same assumption led to additional incoherencies earlier in the history of philosophy, and in psychology and physics. Neutral monists such as James [1912] and Mach [1897] utilized 'sense' as a presumed neutral, non-intentional ground from which to pass either to "matter" or to "mind" according to the nature of the problems they sought to address. This same assumption, that sensation is neutral, also appears in Stout's [1901] theories.[51]

3.8.1. The Presumed Neutrality of Sensation

What is actually *present* in sensation is far from clear. As argued by Russell, our concept of space is one very good example of the ambiguity involved in the concept 'sense'. Matter is often defined as "what is in space." Indeed, sensations are material in that they are stimulations of our sensory receptors or nerve endings, parts of our body which "occupy space." But as soon as the concept 'space' is examined, we find it is highly ambiguous and uncertain. Mathematicians have long known of a multiplicity of possible spaces and have shown many logical schemas that fit the same empirical facts.

Indeed, our subjective experience of colors, for example, has its own subjective "space" and it is clearly not the same "space" (which is a mathematical and scientific construct) that our bodies occupy. The visual experience of colors in the preattentive phase has a different "space" than the "space" our bodies move around in. We know that the posterior cortex contains multiple representations of space which guide movements, grasp, reaches, feeding, and saccades (rapid, intermittent eye movement). Additionally, some of those representations of space in the posterior cortex are mapped on forms of egocentric frames of reference, *e.g.* retinotopic, head centered, body centered. Some map space that is near, and others map space that is far. On the basis of just the concept of space alone, Quine's underlying assumption, that sensation (the senses in general) are cognitively *neutral*, cannot be used as a basis on which to build a theory of knowledge or an understanding of human awareness or consciousness.

A positive feature of Quine's theory is his rejection of at least *one kind* of mentalism. That is, he rejects the view that when we know an object there is in our mind an *idea* (or representation) of the object. Possession of the idea constitutes one's knowledge of the object. But he goes further than this, also rejecting all "abstract ideas" such as universals, propositions, and all intentional phenomena. Somewhat comparable to the earlier neutral monists, he holds that the "mental" (including the logical) and the physical are ultimately comprehensible *only in terms of the same "stuff,"* our stimulatory sensations, which are physical. But this is built on fallacious assumptions, including the assumption that sensation is cognitively neutral.

But there are additional arguments against this and against his view that there is no direct awareness or knowing of an object (*i.e.* not mediated by sentence meaning derived from stimulatory patterns of a community of speakers of the same language). He holds that there is nothing cognitive in the mere presence of an object to the mind. Recall his rejection of "sense data" and his adoption of words to which any given speaker of the language can agree, as the basis for knowledge. Quine requires that there be a *prior* system of interrelated *linguistic* stimulatory patterns of meaning [of a language speaking community] experienced by me before I can *cognitively* experience even one thing. Things become part of my cognitive experience only by virtue of their relations to the *linguistic* stimulatory relations or patterns of a language community. In sum, a person or mind with only one experience, outside a linguistic community, is a logical impossibility. In addition to numerous experiments of new-born babies who have no linguistic community, one of Russell's thought experiments might be used to refute this.

> If I see a particular patch of colour, and then immediately shut my eyes, it is at least possible to suppose that the patch of colour continues to exist while my eyes are shut. . .It seems to me possible to imagine a mind existing for only a fraction of a second, seeing the red, and ceasing to exist before having any other experience.[52]

That momentary acquaintance with the patch of red would seem to deserve to be called cognitive. But even if we reject that it is, there is ample experimental evidence to refute the claim or assumption that sensation is cognitively "neutral" in some sense. In later analyses of the preattentive phase and attention system of our neurological system, I will present experimental findings showing that sensation is not cognitively neutral.

3.8.2. The Problem of Selectiveness of Experience

An adjunct to this problem with Quine's theory is the inability of his theory to explain the *selectiveness* of cognitive experience (as opposed to the mere passive *receptiveness* of sensations). There is nothing there with which to explain it. That is, to a given person at a given moment, there is an object of *attention* (or at most a small number of objects of attention). That object may be given a proper name such as 'this'. As we noted above, before the learning process (as Quine's process of "induction") begins, such selection is already operative, and is *presupposed* in both the process of learning and the process of coming to know in Quine's theory. If such a principle of selection were not operative, where there are not specifically *mental* facts, all experience would be completely diffused, alike, and there would be no distinctions among things. Though this principle is assumed in his theory, nowhere does he either recognize it or attempt to explain it. *A complete theory of human knowing, of natural intelligence, must provide an account of this principle of selection.*

3.8.3. Conflation of Belief and Sensation

Furthermore, the tendency to conflate belief and sensation obscures the differences with respect to what one is thinking or what is before the mind, as well as questions of incorrigibility, fact and error. If I believe that yesterday was Tuesday, there are no sensations equivalent to or identical with the objective content of that belief. But what is believed is that there is an entity answering to a certain description. If I believe that yesterday was Tuesday when in fact it was Monday, then "that yesterday was Tuesday" is not a fact. If I believe that I saw a unicorn, and that unicorns exists, my hallucination is a fact, not an error. My *judgment* based on that hallucination is erroneous. We cannot find in the world any entity, such as a false proposition, corresponding to this belief.

3.8.4. The Rejection of Abstract Objects or Universals

Moreover, if we hold with Quine that there are no abstract objects or universals independent of stimulatory patterns of meaning of a language community, *we would have to hold that '3+3=6' is an entity which exists only at the time some group is disposed to believe it.* On Quine's theory, for '3+3=6' to be "true" or "adequate," that is, supporting our habits and expectations, '3+3=6' must have a relation to the extraneous temporal particulars of stimulatory patterns of language speakers. But there is no temporal particular which is a constituent of this proposition, nor are there stimulatory patterns of a community of language speakers that match up with '3+3=6'.

A final objection to Quine's appeal to dispositions and disposition terms must be raised. Recall that he defines 'belief' as "a disposition to respond in certain ways when the appropriate issue arises." But the concept 'disposition', used by Quine to define a concept central to his theory of knowledge, can in some respects be seen as *pseudo*-scientific. For ultimately, we must ask what can it be used to explain? To claim that one believes *x* because one "has a disposition to respond in certain ways when the appropriate issue arises" is to beg the question. That is, on the basis of *what* do we confirm the relation between empirical evidence (instances of the person's responses in certain ways when the appropriate issue arises) and the *hypothetical attribution* of a belief (that is, a *disposition*) to them?

To define belief in terms of dispositions so as to provide a causal physical explanation of 'belief' in effect *begs the very question at issue*: it is to introduce a *pseudo* concept which masquerades as an explanation. This entire approach entails a much greater problem of the necessary and sufficient conditions of valid projection (confirmation), the problem with so-called disposition terms. The problem of confirmation or valid projection *is* the problem of defining a certain relationship between evidence or base cases and hypotheses and predictions.[53] Though efforts persist, there is no solution to what amounts to the problem of induction generally, and also the problem of counterfactual conditionals. The point here is that Quine's appeal to dispositions not only begs questions best avoided, it *explains* nothing, and actually introduces even more problems.

As possibly the strongest and most forceful proponent of naturalism as the way to explain and resolve issues and problems about human knowledge, Quine's theory fails. His methodological approach fails to explain human knowing and is far from any approach that should be used to understand natural intelligence in general. If we assume that one can only learn by means of direct conditioning to stimuli, by virtue of the arguments presented thus far, that assumption is wrongheaded and stems largely from an unacceptable naturalist approach to theory of knowledge. The evidence shows that the learner brings to the coming to know process the cognitive capacities and structures to make meaningful (not random, nonsensical) exhibitions of his or her experience of the world, of his or her own cognitive development.

For Quine and many other naturalists, human knowing is made of *linguistic* reports, that is, observation sentences and analytical hypotheses. The naturalist/gradualist argument results in incoherent attempts to explain language learning itself, including mathematics learning. In part, this is based on the fact that such explanations replace the traditional immediate awareness, acquaintance, basic, or self-evident knowledge or knowing by a more problematic appeal to "inner" experience or "sense of similarity" of "subjectively natural kinds," that are even more objectionable (and certainly more mysterious). Proof of such incoherencies would thus seem to indicate the necessity to turn attention toward *states of affairs* for exhibition of epistemic significance and away from the biased dependency on the propositional content of language statements.

---

[1] Peter Geach, *Mental Acts: Their Content and Their Objects*, Humanities Press, 1957.

[2] W.V.O. Quine, "Norms and Aims," in *The Pursuit of Truth*, Harvard University Press, 1990. Quine later modified his views as presented in 1969.

[3] Goldman, Alvin, *Liaisons: Philosophy Meets the Cognitive and Social Sciences*, MIT Press, 1992.

[4] Stich, Stephen, and Richard Nisbett, "Justification and the Psychology of Human Reasoning," in *Philosophy of Science*, Vol. 47, pp. 188-202.

[5] Kornblith, Hilary, "In Defense of a Naturalized Epistemology," in *The Blackwell Guide to Epistemology*, John Greco and Ernest Sosa (eds), Blackwell, 1999, pp. 158-169.

[6] Gilbert Harmon, *Thought*, Princeton University Press, 1977.

[7] Coherence theorists are those emphasizing the interrelatedness of language statements. See Keith Lehrer, *Knowledge*, Oxford University Press, 1974.

[8] Giacomo Rizzolatti and Michael A. Arbib, "Language Within Our Grasp," in *Trends in Neuroscience,* Volume 21, number 5, 1998, pp. 188-194.

[9] E. Steiner, *Methodology of Theory Building*, Educology Research Associates, 1988.

[10] Quine, 1978, p. 10.

[11] Ibid., p. 16.

[12] Ibid., p. 22, emphasis mine.

[13] Ibid., p. 33. Also see his *Theories and Things*, 1981.

[14] Graves, Katz, *et al.*, "Tacit Knowledge," in *The Journal of Philosophy*, Vol. LXX, No. 11, June 7,1973.

[15] The notion of 'natural language' used here will follow the meaning of the term found in Nordenstam's [1972] chapter on "The Artificial Language Approach," in *Empiricisim and the Analytic-Synthetic Distinction*, p. 61 ff. Generally, it is taken to mean languages which are historically given with no explicit rules laid down which governed their use from the start, rather because they continually change, the rules must be found out. This is in contrast to artificial languages which are essentially simple, with rules explicitly set forth. There is no question of rules being right or wrong in artificial languages except with reference to the

purpose at hand, while with natural languages, the rules are taken to be descriptive of that language.

[16] W.V.O. Quine, *Ontological Relativity and Other Essays*, Columbia University Press, 1969, p. 26, brackets mine.

[17] Quine, "Grades of Theoreticity," in *Experience and Theory*, Foster and Swanson (eds.), University of Massachusetts Press, 1970, pp. 4-7, emphasis mine.

[18] Quine, *Ontological Relativity and Other Essays,* 1969, p. 27.

[19] Quine, 1969, p. 27, emphasis mine.

[20] Quine, 1970, p. 13ff.

[21] Quine, 1969, p. 29, emphasis mine.

[22] Quine, 1970, p. 7.

[23] Quine, 1970, p. 6.

[24] Peter Geach, *Mental Acts: Their Content and Their Objects*, New York, Humanities Press, 1971, p. 26 [information in brackets mine].

[25] Among many other research studies, see Adam K. Anderson and Elizabeth A. Philps, "In Neuroscience First, Researchers at Yale and NYU Pinpoint the Part of the Brain that Allows Emotional Significance to Heighten Perception," *Nature*, Vol. 411, May 17, 2001, pp. 305-309. Summary in *Science Daily Magazine*, 18 May 2001.

[26] Ibid.

[27] Giacomo Rizzolatti and Michael A. Arbib, "Language Within Our Grasp," in *Trends in Neuroscience*, Volume 21, issue 5, 1998, pp. 188-194. Also see: "Monkey Do, Monkey See. . Pre-Human Say?" summary in *Science Daily Magazine*, August 20, 1998.

[28] *Science Daily Magazine*, Editors, "Computer Program Trained to Read Faces Developed by Salk Team, " Summary, March 22, 1999.

[29] Francis Bacon, *The New Organon and Related Writings*, F. Anderson, (ed.), Liberal Arts Press, 1960.

[30] M. C. Bradley, "Comments and Criticism: How Never to Know What You Mean," in *The Journal of Philosophy*, Vol. LXVI, No. 5, March 13, 1969.

[31] Quine, 1969, p. 122, emphasis mine.

[32] See Nelson Goodman, *Fact, Fiction, and Forecast*, Bobbs-Merrill Publishing Company, 1973.

[33] Quine, 1970, p. 16.

[34] Thomas S. Kuhn, *The Structure of Scientific Revolutions*, Second Edition, International Encyclopedia of Unified Science, University of Chicago Press, 1970, p. 193, emphasis mine.

[35] *Science Daily Magazine*, (eds.), "New Approach to Imaging Separates Thought From Perception," Summary, October 26, 1999.

[36] Again, see Adam K. Anderson and Elizabeth A. Philps, "In Neuroscience First, Researchers at Yale and NYU Pinpoint the Part of the Brain that Allows Emotional Significance to Heighten Perception," *Nature*, Vol. 411, May 17, 2001, pp. 305-309.

[37] Edwin Martin, "The Intentionality of Observation," in *Canadian Journal of Philosophy*, Vol. III, Number 1, September, 1973.

[38] Quine, 1978, pp. 23-24.

[39] For starters, again, see Anderson, Adam K., and Elizabeth A. Phelps (2001). "Lesions of the human amygdala impair enhanced perception of emotionally salient events," in *Nature*, Vol. 411, 17 May, pp. 305-309.

[40] Edwin Martin, "The Intentionality of Observation," in *Canadian Journal of Philosophy*, Vol. III, Number 1, September, 1973, pp. 121-129.

[41] Also see "Study Finds New Way That Brain Detects Motion," in *Nature*, April 12, 2001. The discussion of how the brain measures self-motion to determine how quickly we are hurtling

toward something, gives an excellent neurological foundation for apparently simple (but actually very complex) tasks such as knowing how hard to hit the brake when we're driving a vehicle.

[42] Recall that a language community is composed of speakers who already know the language.

[43]This is due to what he claims are inadequacies with the concept of synonymy of meaning.

[44] Quine, *Word and Object*, MIT Press, 1960, p. 67.

[45] Peter Geach, 1971, p. 43.

[46]For example, see Kessen [1965], Bowers [1965], Bruner [1966], Repp [2001] and Anderson and Phelps [2001] to name only a very few.

[47]For an interesting comparison of normal infant face recognition and autistic children, see Geraldine Dawson in G. Dawson and K. Fischer, (eds.), *Human Behavior and the Developing Brain*. New York: Guilford, 1994.

[48] William Kessen, 1966, p. 14.

[49] T.G.R. Bower, "The Visual World of Infants," in *Perception: Mechanisms and Models*, Readings from Scientific American, San Francisco, W.H. Freeman and Co., 1972, pp. 349-357.

[50] Hector Neri Castañeda, "Philosophy as a Science and as a Worldview," in *The Institution of Philosophy*, Avner Cohen and Carcelo Dascal, (eds.), Nous Publications, Indiana University, Bloomington, Indiana, 1990.

[51]George Frederick Stout, *A Manual of Psychology*, 2nd edition, London: W.B. Clive, University Tutorial Press, 1901. Also see Bertrand Russell, 1984, pp. 21n-22.

[52] Bertrand Russell, 1984, p. 23.

[53] Nelson Goodman, *Fact, Fiction, and Forecast*, Bobbs-Merrill Publishing Company, 1973.

4. WHAT DOES THE EVIDENCE SHOW?

". . .we form our ideas also of those objects on the basis of something else which is immediately given."
Kurt Gödel[1]

This chapter will specifically address this issue: What is the evidence for immediate awareness? What is a scientifically acceptable definition of 'immediate awareness'? What is the evidence that it is cognitive? Do we actually select objects as unique, as *sui generis*? If so, how do we actually do it? What does the evidence show? In light of strong arguments that our brains are classification "machines," wetware containing algorithms that classify, what arguments are there to show that we do anything other than this?

Any theory attempting to establish that human beings have a cognitive *immediate* awareness relation with a unique object must establish several things: (1) It must establish that there is some level of awareness that *is* cognitive and *is not* mediated by propositional statements, symbols, or linguistic units of any kind; and (2) It must establish that there are *sui generis*, unique, objects in that relation of awareness. Furthermore, the theory must present arguments showing that prior definitions of immediate awareness are inadequate. This section will specifically address both empirical and logical evidence and arguments for cognitive immediate awareness, based largely on experimental neurophysical, cognitive, and psychological research.

Moreover, I will sort out the hierarchy of primitive relations to obtain a more graphic view of how they are ordered relative to one another.

4.1. Problems with Subjective Definitions of Awareness

Most earlier attempts by theorists or researchers to address the problem of awareness in general tended to rely upon introspective reports of their own inner, subjective experiences. They also identified attention as the "starting point" for intentional, cognitive activity. We saw this above in Russell's theory and it can also be found in James' theory and others. Though subjective reports of one's internal states should not necessarily be entirely thrown out of any experimental inquiry involving the mind, the problem with such attempts is that subjects' subjective reports of their inner experience may very well be influenced by bias. Any given subject may claim to be unaware of a stimulus unless they are completely comfident in their response. Alternatively, a subject may claim to be aware on the basis of just about any sensation. Individual subjects may tend to determine whether or not they aware on the basis of their own private criteria for awareness. Thus such reports cannot be used to precisely define awareness, including immediate awareness.

Subject bias can also be found even in experimental studies on awareness without introspective reports. In sensory discrimination tasks, for example, there is evidence that subjects are systematically underconfident, hence they may systematically claim *not* to see stimuli that they have partially or even entirely seen.[2] Moreover, with objective definitions based on correct *versus* incorrect identifications by the subject, subjects making an incorrect identification may nonetheless still have some awareness of the stimuli. Even with objective definitions based on chance and greater than chance performance, issues of whether perception of a stimulus can occur without awareness will not be resolved because they are insensitive to subjects' phenomenal experience. There are other approaches as well with similar or even more complicated problems.

The most promising approach to objectively measure awareness appears to be offered by Kunimoto, *et al.*[3] Their proposal is to measure awareness in terms of subjects' ability to discriminate between correct and incorrect responses using a metric provided by Signal Detection Theory (SDT). 'Awareness' is operationally defined as follows: a subject is aware if and only if confidence is related to accuracy (with the metric greater than zero). The approach uses both subjective reports for assessing awareness by analyzing confidence reports with techniques developed in SDT to eliminate response bias. This operationally defined concept of awareness still ties awareness to subject reports of their own inner states. Thus, though this method may overlap in some ways with our concerns, it does not directly address immediate awareness. It is apparently more addressed to "awareness that" than immediate awareness.

One useful suggestion by Kunimoto, *et al.*,[4] however, is that the general concept should not be viewed in terms of two mutually exclusive states, awareness or unawareness. It should be viewed as a continuum of states ranging from unaware through an infinite number of partially aware states, to complete awareness. However, they have not distinguished between "awareness that" such and such is the case [tying awareness to "that" clauses or linguistic reports] and "immediate" awareness which is not tied to such reports, though their concern is with subliminal awareness. This distinction should be factored into any continuum, with a clear map showing where the two categories lay on it. Sorting a hierarchy of "sheets" of primitive relations of awareness, including those of immediate awareness, and showing where they lay on the continuum, poses a challenge that we will address below.

4.2. Neurophysical Experiments

Early neurophysical experiments by Libet[5] asked the following question: How elaborate must spatiotemporal neuronal activity be for a subject to *consciously* perceive it? He sought to establish a threshold for awareness, below which the subject is unaware, while above it the subject is aware. In those experiments, 'awareness,' 'conscious perception' (or conscious 'awareness') is interpreted by Libet as "awareness *that*" or "consciousness that" such and such is the case. That meant that the subject could tell or otherwise indicate to Libet that he or she felt the stimulus. So though his experiments were not directly related to the immediate awareness of concern to us,[6] nonetheless, his findings are a good place to start looking at the evidence.

Libet[7] conducted experiments in which he employed gentle electrical stimulation to the cerebral cortex of subjects and to the skin of the hands of those same subjects. He found that a brief *repetitive* stimulation of the sensory cortex was far more effective in evoking a perceptual experience than was a *single* stimulus. Varying the train of the stimulus, it was found that there can be a conscious experience only where there has been time, up to 0.5 s, for an elaboration of the spatiotemporal patterns in the neuronal machinery of the sensory cortex. In contrast, he also found that a single weak cutaneous stimulus could be perceived just as well as a train.

Through continued experimentation, Libet produced some surprising findings. He showed that though a single skin stimulus requires up to 0.5 s of cortical activity before it can be experienced, under certain experimental conditions, it is antedated to the initial evoked response of the cortex. That is, under experimental conditions, it was shown that there is a kind of "backward masking" and "antedating" of the stimulus. But he could give no account of the mechanism that does this.

Eccles later proposed an hypothesis which claimed that it is the *self-conscious* mind, not the neural machinery of the brain, that acts to select from the multitude of active centers at the highest level of brain activity, according to attention, to give unity to the most transient experiences. He states:

> The self-conscious mind is actively engaged in reading out from the multitude of active centres at the highest level of brain activity. . .The self-conscious mind selects from these centres according to attention . . Thus we propose that the self-conscious mind exercises a superior interpretative and controlling role upon the neural events. [8]

He also advanced this hypothesis to account for apparent paradoxical findings such as Libet's. A central part of his hypothesis to note is the implied distinction between the hypothesized *self-conscious mind* and the mechanism of *attention*. Note also that 'self-conscious' appears to be "consciousness *that*" such and such is the case. If I am reading his hypothesis correctly, there are two separate things, the self-conscious mind and the mechanism of attention, with the hypothesized self-conscious mind somehow in control of attention. But a point to note is that he has actually recognized that there is a deeper level of awareness than the "awareness that" which Libet's experiments addressed. Libet's experiments depended upon subjects' ability to tell him whether or not they felt something when he applied a stimulus. It was the apparent incongruities or paradoxes in his findings, the "backward masking" and "antedating" of stimuli, that reveals a more primitive level of awareness, beneath the "awareness that" level.

But I have introduced Eccles' hypothesis here largely as a kind of "jumping off" point to analyze related issues about the brain and its active selection. We should review some basic distinctions about the human brain, specifically distinctions between the preattentive phase and attention system, and some assumptions.

4.3. Cortical Information, the Preattentive and Attentive Phases

Living organisms such as humans have a sensory apparatus capable of identifying stimuli by means of a filter consisting of signals generated by the apparatus itself. They encounter a world of color, sound, texture, shapes and contours that the sensory apparatus produces and selects by means of its filter. Some theorists, confusing symbols for the things symbolized, fall into subtle nominalist traps when they describe what happens this way:

> With these filters and analyzers, the sensory systems "invented" an entirely new form of information: Instead of physical properties that cannot be transferred to sensory channels, a *representation* of them was selected and produced, namely, the filtered sense qualities. Such a representation is also referred to as a "symbol"; therefore, one may refer to sense qualities as elements or signs of symbolic information.[9]

It is best to remember that we are the ones who *represent* information as symbolic; it is not symbolic information that the filters and analyzers of the sensory system are handling. Such descriptions appear to be used metaphorically, not literally. But neurological sensory data and impulses are not themselves symbols and the use of metaphors in such descriptions to describe them that way can lead to problems. It is the confusion of an object represented with a representation of the object that leads to such talk and to further confusion between levels of awareness, as well as to wholesale fallacious inferences and theorizing that collapses the levels and begs questions at issue. Some of these problems, no doubt stemming from an overwhelming nominalism, can be found in the following standard descriptions.[10]

Organisms store and analyze information in the cortical network, centralizing controls in the reticulo-thalamo-cortical (RTC) system. The complex interactions and transactions with the environment are enabled by the neocortical network, whose primary and secondary sensory areas represent the peripheral sensory receptor system in the cortex. These areas continue the functions of analysis and filtering of information from the environment. As part of the sensory system's filtering function, the visual system analyzes differences in light, colors, movement, shapes and contours. It is important to note that the filtering function is the means by which the sense qualities are *selected* before the act of seeing can take place. The cortical sensory detectors are the carrier of code for the sense qualities which have to be decoded into information in order to be meaningful.

Preattentive analysis precedes the first storage of information and conscious perception, having a latency period of about 60 ms. Signals are transmitted to the sensory fields of the cortex. During the preattentive phase, the RTC and the stimulus excite primary arousal of the activation system itself and the sensory fields. The body and its senses become aligned with the stimulus *via* the sensomotoric paths of the reticular brain stem. The function of the sensory system during the preattentive phase, including the sensory fields of the cortex, is to analyze stimuli so that the sensory system can filter the stimuli and align the filtered sense qualities with the stimulus.

According to experts,[11] preattentive orientation proceeds *subconsciously* (which appears to be interpreted as the absence of "consciousness *that*" such and such is the case) at the level of the nervous system. It is only when sensory perception is attained that attention can then focus upon information as an object with which it can operate. Only when this level is reached does preattention make the transition to the *conscious* attention of a cognitive system. And this appears to be the line between *non-cognitive* neurophysical activity and *cognitive* neurophysical activity, according to these experts. Only "consciousness that" or "awareness that" is held to be cognitive. Any activity below this is held to be non-cognitive.

But one must raise issues with the description given of the "preattentive phase" as "pre-" and with the use of the concepts 'conscious' and 'attention', and the implied distinction between the cognitive and non-cognitive. It is evident that during the preattentive phase, the body and its senses become aligned with the stimulus by way of the paths of the reticular brain stem. Excitation levels of certain areas are raised in preparation for uptake and processing of sensory signals. The task of the sensory system in the preattentive phase is to analyze stimuli so that the sensory system can filter the stimuli and align the filtered sense qualities with the stimulus. In other words, during the preattentive phase, the organism is *already* making preparations and *aligning its senses* with some stimulus. Logically, this implies that the organism is *already directing itself* in neurological ways to attend to some stimulus that it has already *in some more primitive sense* selected to align itself with. It has to have made such a selection since any given stimulus would be in an environment filled with possibly an infinite number of stimuli from which to select.

So one must ask, "Just was is so *pre-* about the *preattentive* phase?" Obviously, it is intended to mean that conscious perception of an object has not yet taken place. But the organism appears to be *already* attending to a stimulus that it has already selected out of a possible infinite number from which to select. The preattentive phase is said to precede conscious sensation in the activation of attention. The use of the term 'conscious' here is tied to "awareness that" such and such is the case. And it is in the attention system combined with the activation system that, so it is claimed, cognition occurs.

But this is not only conceptually confused, it is also fallacious. It is conceptually confused for at least those reasons pointed out above, and it is fallacious because of the evidence of *intentional*, cognitive activity even during the preattentive phase, the phase that is otherwise described as being "without awareness." Keep in mind that in the above description, the term 'awareness' means "awareness *that.*"

4.4. The Primitives of the Preattentive Phase

We should look more closely at exactly what is going on during the preattentive phase and what function that phase serves. The preattentive phase processes features of objects, but not objects. It processes features without attention and without awareness *that* one is processing them. Tasks that can be performed on large multi-element displays in less than 200 to 250msec are considered preattentive. In many experiments, subjects could perform search tasks in time less than 200msec. This clearly suggests that the information is parallel-processed.

Also, the preattentive phase processes single features. Experiments by Wolfe, *et al.*[12] have shown that "we can know preattentively that an object has the attributes 'red' and 'vertical' and yet have no idea if any part of the object is red *and* vertical."[13] In general, to process a conjunction of features, including two or more of the same features appears to require attention.[14] But it is that preattentive processing of primitive feature information that prepares the deployment of attention, when perception occurs, and taken to be the beginning of cognition. That preattentive processing exists to *direct* attention to the locations of interesting objects in the visual field. It follows that the directional activity of preattentive processing must be intentional.

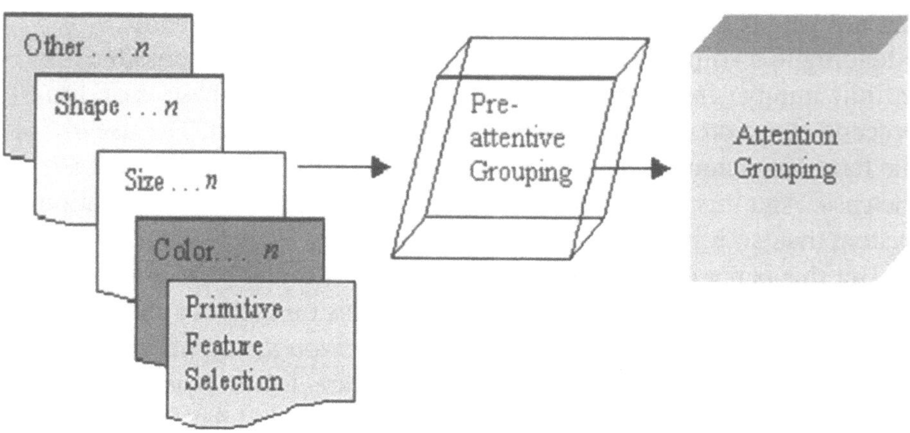

Figure FOUR-1. Preattentive Feature Process

Approximately eight to ten basic primitive features used in the preattentive phase have been identified, further confirming earlier arguments that what is *actually present* in sensation is not entirely clear. But what is clear is that sensation is not cognitively "neutral," as Quine and other naturalists assume. There is clear evidence of intentional activity during the preattentive processing of features and directing of attention. The primitives identified that are used during the preattentive phase (but of course not necessarily in each case) minimally consist of the following: color, orientation, motion, size, curvature, depth, vernier offset (small departures from the colinearity of two line segments), gloss and, perhaps, intersection and spatial position/phase.

As Wolfe notes,[15] there may be a few other local shape primitives to be discovered because the primitives of preattentive shape processing are not entirely known. Recalling Russell's and Gödel's comments about "the given" and acquaintance with universals (permitting us to have knowledge of mathematics and logic), as well as our earlier objections to Quine's assumption of the neutrality of sensation, the problem is a lack of a widely agreed upon understanding of the layout of "shape space." Shape or form appears to be the most problematical primitive feature in the preattentive phase. For example, simple color space is a two-dimensional plane or it could be three-dimensional if the surface has luminance. As Wolfe notes, it is not clear what the "axes" of shape space might be. But preattentive processing of "shape space," whatever we take that to be, enables us to then make sense of objects we attend to, and to make sense of a whole lot of other properties of things, including motion.

Moreover, there are differences in how each of the primitives is actually processed in the preattentive phase. However they are in fact processed, they are used to intentionally guide attention to some object. Assuming that organisms are not designed to do too many wasteful things, such as processing a lot of features that are not necessary to guide attention, it is probably safe to assume that any processing during the preattentive phase is usually efficient and done for the sake of *successfully* guiding attention to some object. There is a rather long list of research[16] that essentially sorts out two ways that processing occurs:

either bottom-up, or top-down. We will briefly look at bottom-up processing.

We can sort out a number of hypothetical test situations: (1) In a situation with a target feature (such as color) that is sufficiently different from distractors, efficient search occurs even without identity of the target; (2) In a situation where a series of items are grouped by feature, attention is shifted to the border where the feature changes; as the size of the group increases, some feature searches get easier. In both sets of situations, the phenomenon of "pop-out" occurs. In (1), the pop-out is feature search for an unusual item; in (2) there is a continuum of pop-out.

Wolfe considers a variation on what is termed a "singleton" search. This is a search in which a single target is presented among homogeneous distractors and differs from those distractors by a single basic feature. Preattentive processing of the unique item causes attention to be deployed to that item so it is examined before any distractors are examined.[17] In each of these test situations, preattentive processing of information is intentional. Further, the processing is done by feature selection as unique objects, not based on similarity of a feature with other features. All this occurs, by the way, in the absence of attention. It occurs in the absence of any "awareness that" it is occurring.

Of course, as Wolfe points out, searches for stimuli are usually not searches defined by a single feature. We do not usually look for "red," but look for an apple that is a conjunction of a set of features such as red, curved, shiny, and being the size of an apple. Feature integration theory and on-going research are tackling the problem of how conjunction searches in the preattentive phase occur, a matter I will not pursue here.

4.5. Evidence for Cognitive Immediate Awareness

Evidence for *cognitive immediate* awareness ("awareness of" in contrast to "awareness that") activity during the *preattentive phase* has been empirically shown or strongly suggested in a variety of research studies.[18] I cite only a few here that will be relevant for discussion.

In experiments testing rapid visual categorization in the absence of awareness, Van Rullen, *et al.*,[19] subjects were asked to respond to masked and unmasked natural scenes when they contained an animal. In addition, subjects rated their confidence in perceiving the contents of each masked image. For a majority of the scenes, masking effectively prevented awareness of the stimuli, as indicated by the fact that confidence ratings did not predict categorization accuracy. For the same scenes, however, subjects responded *significantly above chance level* to the presence of animals.

In addition, in the same experiments, motor responses started to reflect *correct* categorizations at the same time for masked and unmasked stimuli, indicating that early responses in "normal" (unmasked) visual categorization probably also rely on the first milliseconds of stimulation. Similar results were obtained with simpler displays for which stimulus and mask contrast could be controlled. In that case the earliest motor responses to "perceived" and "unperceived" targets showed virtually identical distributions. According to the researchers, these experiments showed that information about the first milliseconds of visual stimulation can propagate throughout the visual system, unaffected by later changes, and determine behavior *even when it is not (or not yet) available to consciousness.* Again, 'consciousness' here refers to "consciousness *that.*"

In a related study, Kunimoto, *et al.*,[20] using their method described earlier, conducted four subliminal perception experiments using the relationship between confidence and accuracy to assess awareness. Subjects discriminated among stimuli and indicated their confidence in each discrimination response. Subjects were classified as aware of the stimuli if their confidence judgments predicted accuracy and were classified as unaware if they did not. In the first experiment, findings indicated that subjects' claims that they are "just guessing" should not be accepted as sufficient evidence that they are completely unaware of the stimuli. Experiments 2-4 tested directly for subliminal perception by comparing the minimum exposure duration needed for better than chance discrimination performance against the minimum needed for confidence to predict accuracy. The latter durations were slightly but significantly longer, suggesting that under certain circumstances people can make perceptual discriminations *even though the information that was used to make those discriminations is not consciously available.* 'Consciously' again means "consciousness *that*."

Repp[21] conducted research on finger-tapping which revealed an internal mechanism which guides motor actions in response to subliminal changes in stimuli. Through a total of five experiments, subjects were assessed in terms of sensorimotor coordination, phase correction, timing adjustment of a repetitive motor activity to maintain synchrony or some other intended temporal relation with an external sequence of events. They were also tested in terms of phase resetting which is a more dramatic timing adjustment that immediately restores synchrony after a large synchronization error. In each test, subjects correctly altered their motor actions in response to subliminal changes in stimuli even without a conscious perception of change. Repp concluded that the brain agent guiding the motor behavior is below the perceptual threshold. At some level, the brain is much more sensitive to timing information than the results of previous psychophysical experiments suggest. This precise timing information seems to be used in the control of actions, *without awareness* ("awareness that").

In Colombo, *et al.*,[22] researchers tested visual search asymmetries in 3- and 4-month old infants indicative of a *preattentive* phase. Thirty-two infants from each age group were presented with 2 visual arrays to the left and right of midline. The stimuli were constructed of feature-positive and feature-absent arrays, each paired with a corresponding homogeneous array in which no discrepant element was embedded. The visual fixations of infants were measured, showing a "pop-out" effect for feature-present stimuli in both age groups. The results were similar to but not as strong as results found for adults. As the researchers note, the findings may reflect limitations of infant visual search, the methodology used to assess it, or the difference in the size of the effect between adults and infants. The findings also show evidence of visual quality selection in the preattentive phase for infants.

In Näätänen, *et al.*,[23] tests were conducted with multiple simultaneously active sources of seemingly chaotic composite signals, with overlapping temporal and spectral acoustic properties, impinging on subjects' ears. In spite of the chaotic composite signals with overlapping temporal and spectral acoustic properties, the subjects' perception is an orderly "auditory scene" that is organized according to sources and auditory events. This allows them to select messages easily and recognize familiar sound patterns, and to distinguish deviant or novel sound patterns. The data suggest that subjects' ability to organize such impinging signals is based on a kind of "sensory intelligence" [sic] in the auditory cortex. "Even higher cognitive processes than previously thought, such as those that organize the auditory input, extract the common invariant patterns shared by a number of acoustically varying sounds, or anticipate the auditory events of the immediate future, occur at the level of sensory cortex (even when attention is not directed towards the sensory input)."

Some studies on "blindsight" or "numbsense" convey how some persons who are conventionally blind or insensible by objective measures can nonetheless discriminate visual or tactual test stimuli correctly with near-perfect accuracy. These patients will insist that they can't "see" or "feel" anything despite objective evidence to the contrary, demonstrating a level of awareness I refer to as "awareness of" not reducible to the subject's "awareness that."[24] Subject actual responses *correlated negatively* with their verbal reports.

Similar studies conducted decades earlier showed that subjects presented with a series of nonsense syllables, who were then subjected to mild electric shocks at the sight of certain syllables, soon showed symptoms of anticipating the shock at the sight of "shock syllables." Yet, on questioning, they could not identify the syllables. The subjects had come to know when to expect a shock but could not tell what made them expect it. These findings are similar to other experiments which showed they knew or could identify persons by signs they could not tell.[25] The findings also seem to suggest that subjects also knew patterns of *timing* associated with "shock syllables" by signs they could not tell.

Finally, experiments in human perception show that in spite of "noise" in images and gaps in contours caused by light intensity variations and occlusions, human perception is able to account for these by using an intrinsic process of line completion and grouping of parts into whole entities. There is evidence that this entire process is purely preattentive without any top-down (know*ledge* or "awareness that") influences.[26] It is this very complexity of primitive relations of immediate awareness in the preattentive phase that poses such enormous obstacles in building artificial systems that can detect and recognize objects, including human motion.

The above experiments involve most of the sensory and the somatosensory-motor system and very large numbers of primitives in the preattentive phase and attention system. They show *cognitive* immediate awareness of objects that is not mediated by linguistic units. In fact, certain of the experiments, as in the blindsight and numbsense experiments, showed that subjects' correct responses correlated negatively with their own verbal reports. Moreover, the activity of preattentive feature selection cannot be the activity of classification, though researchers may describe it, in some metaphorical sense, as classification. It does not proceed by comparing properties of objects based on a principle of similarity. There are, in most preattentive processing, no conjunctions of features, including no conjunctions of like features with which to compare properties. Moreover, preattentive processing takes place in new-born babies, seeing features for the first time.

In the preattentive processing phase alone, we have yet to figure out all the primitives involved and exactly how the process and the interrelations among all the primitives actually work. In the visual system, we have three types of cones that allow us to distinguish between about 2 million colors, but there are probably billions of actual primitive featural relations involved in the preattentive phase and attention system. Again, we do know that however the process works, preattentive processing acts intentionally to deploy attention, the place where actual perception occurs.

The above experiments also involve the visual sense of motion as well as visually guided action. What is called the MT+ region of the inferior temporal sulcus, shown in the figure below, consisting of MT, MSTl and MSTd, has multiple regions specialized in different aspects of motion perception. It is motion perception that extracts the three-dimensional structure of the world, defining the edges and forms of objects. All this involves what are called the "what," "where," and "when" pathways and sharing of information. While some neurons are good at determining the direction in which an object is moving, they cannot identify the object. Some cells in layers of the visual system are sensitive to orientation and also to motion in particular directions. Parts of MT+, MSTl and MSTd, sense when *objects* move; others sense when *you* move. Different patterns of optic flow are produced in your retina when you move in different directions and the neurons in MSTd recognize these different patterns. Analyzing the above experiments, it is fairly easy to see that this enormously complex system is involved in them all.

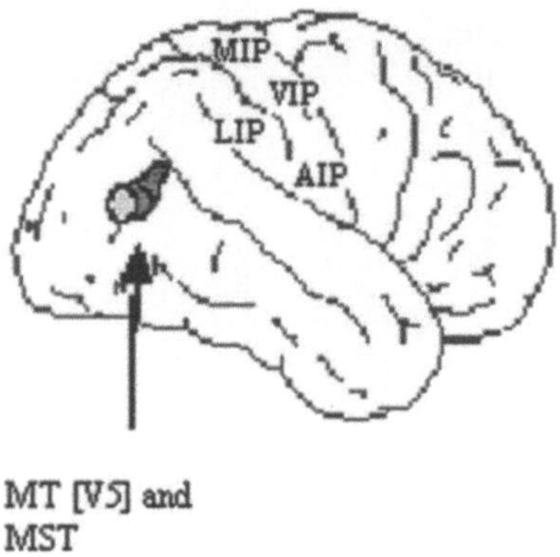

Figure FOUR-2. The MT Region with MST, MIP, VIP, LIP, AIP

Moreover, all the regions of the brain that guide a variety of movements are involved as well. There are multiple representations of space in the posterior cortex that makes all this possible. The LIP (lateral intra parietal) region represents locations of objects that you intend to look at and may reach for. The MIP (medial intra parietal) region represents immediate extra-personal space, which is the space you can reach to; it guides arm movements. The AIP (anterior, intra parietal) region represents the shape information we need in order to grasp objects. And the VIP (ventral intra parietal) region represents the near space used to guide the head, mouth and lips during feeding. This region receives visual and tactile information from the face.[27]

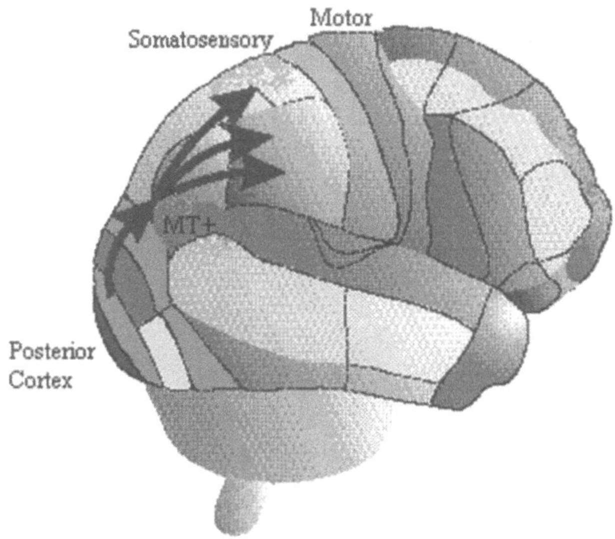

Figure FOUR-3. Brain Showing "Layers" of Motor, Somatosensory, and Posterior Cortex .

With respect to touch, as in the above finger-tapping experiment, our discriminative ability depends on a variety of touch receptors coding millions of stimuli. The somatosensory system includes multiple types of sensation from the body - light touch, pain, pressure, temperature, and joint and muscle position sense, all of which may be involved in highly complex interrelations with one another, in this experiment. Each of these kinds of sensation are carried by different pathways and have different targets in the brain, and each cross one another at different levels.

4.6. Where Do We Enter the Circle of Cognition?

And when do we enter it? No matter where we draw the line between cognition and non-cognition, do we enter the circle the day we're born, soon thereafter, or even before? Tests conducted on newborn babies have shown that they not only already perceive a great deal, they have distinct preferences and soon recognize their own mothers from among other present adult females. Moreover, it has also been found that physiologically normal babies have an intellect that is at work long before language is available to them as a tool. Infants as young as one month can already differentiate between sounds in virtually any language.[28] These babies do not have categories, classes and kinds in their minds when they do all of this. They do not have a set of representations floating around somewhere in their brains that they use to label what they are experiencing.

The above experiments are but a handful out of many more that on the whole reveal, among other things, that we must revise our understanding of the cognitive domain and the place where we enter it. Currently, cognition is viewed as largely starting with the attention system and continuing on to higher levels. But all of these experiments have in common that they showed some deeper level of awareness, below the attention system threshold, that *correctly* affected subjects' overall behavioral responses. The experiments also show, especially in Van Rullen and Kunimoto, that there is in fact a *negative correlation* between subjects' own verbal judgement (*knowledge that*) about their own awareness and their awareness as actually measured in experiments. This shows that there is a non-verbal encoding of task or act-relevant sensory data that is available to the subject at deeper levels of awareness during the preattentive phase. The "circle of cognition" is not entered at the level of the attention system, but before.

It is also evident that the circle of cognition is larger (and deeper) than previously thought. This is so as it pertains to not only vision, but also the psychomotor and entire sensory motor parts of the brain. Näätänen, *et al.*, showed that even in the midst of what has to be described as a noisy, chaotic setting, we have some kind of primitive intelligence in the auditory cortex even when attention is not directed toward the sensory stimuli. But the tension in the research literature brought about by the emphasis upon language or symbol-mediated "knowledge that" or "awareness that" leading us to deny the label "intelligent" to anything other than an exhibition of "knowledge that," is evident even in the title of those experiments conducted by Näätänen, *et al.* In their title, "'Primitive Intelligence' in the Auditory Cortex," the phrase "primitive intelligence" is put in single quotes to imply that, *in spite of evidence to the contrary*, it may not be real.

Recent studies[29] in neuroimaging of cognitive functions in the human parietal cortex have shown that many of the functions it probably serves are components of many cognitive tasks. Though their concerns were with mapping activation in the parietal cortex, they also pinpointed much of the complex activity of a *cognitive* nature revealed by the parietal cortex. As the researchers pointed out, most cognitive tasks involve one or more of the following components: shifting and maintaining attention [and also preattention]; directing eye movements and generating motor plans, explicitly or implicitly; using working memory; coding and transforming space in input or output (*e.g.* arm-centered) coordinates. In effect, they showed that the cognitive mapping of parietal cortex activation generalizes across a broad range of intelligence activity, and that cognitive activity is not limited to language mediated activity.

Furthermore, we cannot correctly conceptualize what's going on here in terms of the "activation" of "dispositional representations" in various parts of the brain. Literally, there are no representations in the brain. There is only what is *present*. The current effort among some theorists to explain what is happening in terms of symbolic (or other kinds) of representations in the mind or brain is an echo of earlier idealist theories of Berkeley and others, with a strong dose of Cartesianism and nominalism thrown in.

4.7. Learning All Over the Nervous System: Multiple Intelligences

The brain is still considered the lofty reservoir of intelligence, of knowing. But recent advances in research on the central nervous system (CNS), consisting of the brain and spinal cord, have led to dramatic changes in views on the subject of intelligence, learning, and the part played by the body in our knowing the world around us and ourselves. Until recently, if sufficiently traumatic, any injury to the spinal cord was thought to be untreatable. But since the early 1980's, new research discoveries and tools in the area of spinal cord regeneration and repair have led to entirely new ways of looking at the spinal cord, the part it plays in our overall intelligence, and whether or not it is treatable after injury. Much of the research on the spinal cord is now directed not just to regeneration, but also to "reeducating" the cord after injury, even when it has been severed from the brain.[30]

There is experimental evidence that suggests that the spinal cord can *learn* on its own. Surprisingly, it may even be able to learn without the brain. Experiments on rats and monkeys to change the size of knee-jerk reflex in exchange for a reward produced anatomical and physiological changes in the spinal cord. Moreover, the results of the training persisted even after the brains of the animals were disconnected from their spinal cords (within humane guidelines), clearly suggesting memory stored in the cord and that learning had taken place *in the cord* itself.

In other research studies, it was shown that the spinal cords of adult animals could be trained after injury. The primary interest of the research was directed to the spinal cord circuitry leading to studies in adult mammals, making the research directly relevant to human subjects. These studies showed that spinally injured cats could *relearn* walking patterns of normal cats. If they were not trained after injury, however, they were less likely to relearn walking patterns. These experiments showed that spinally injured animals learn whatever they are experiencing, and showed the need for task-oriented practice to "teach" the cord to do certain tasks while drawing upon its embedded memory. As researchers pointed out, after spinal cord injury, highly complex changes in the way neurons communicate occur throughout animal nervous systems in order to adjust to preserve old behaviors. The spinal cord acts to make the most of whatever circuitry it has left. Severed spinal cords are sometimes capable of being conditioned on their own, implying, among other things, the ability of the cord to learn from memory within it, and the need to revise not only the concept "learning," but what may also be an outdated notion of "conditioning."[31]

Traditionally, learning was thought to take place in specialized brain centers and stored in designated areas in the form of memory. When activated by some kind of stimulus, the memory store was accessed, and the person "remembered." But that view of learning and memory has undergone something of a change given recent research findings particularly involving spinal cord injuries. We now know that learning is a process that takes place *all over the nervous system*, not just in the brain, depending upon what is being learned.[32] Moreover, we now know that in spite of differences between human spinal cords and lower animals, the need for task-oriented practice is a feature of the nervous system in all species.[33]

Recall that Quine took learning to be a process of induction from sensory stimulations to observation sentences. Learning, in his sense, was tied to linguistic reports and depended upon a learner recognizing their own "inner sense" of "subjectively natural kinds." This is clearly a "heady" sense of learning. Aside from the spurious sense of induction to define learning, one is left wondering how a learner knew beforehand what is similar from what is not. But Quine was more or less following the traditional concept.

Thus not only do we need to revise our understanding of the scope and depth of the cognitive domain and the place where we enter it, we must also revise our understanding of a network of related concepts, including cognition itself, natural intelligence, learning, and conditioning. If the empirical findings and our interpretations of them are correct, the circle of natural intelligence begins with immediate awareness in the preattentive phase. Minimally, natural intelligence involves not just the brain but the entire central nervous system, including both the brain and the spinal cord, in which highly complex, dynamic interactions among primitive relations of the entire sensory and somatosensory-motor systems are involved. Our natural intelligence is of many kinds and exhibited in many ways.

Multiple Intelligences

The need for task-oriented practice in spinal cord injured animals and humans in order to re-learn is fundamental to their recovery of normal or near-normal functions. But this finding of the need for task-oriented practice can be generalized to normal (uninjured) humans in order to develop any kind of intelligence at all. That generalization is supported in part by the research referenced above, especially neuroimaging of cognitive functions in the parietal cortex. It is also supported by research over many decades on the nature of multiple intelligences by Gardner, *et al.*[34]

Gardner, *et al.*, have empirically identified at least six separate and distinct kinds of natural intelligence, basing their research primarily upon neurological, cross-cultural, and psychometric evidence. These kinds of intelligence include *linguistic, musical, logical-mathematical, spatial, bodily-kinaesthetic*, and *personal* intelligences, that involve different parts (sometimes overlapping) of the central nervous system, both brain and spinal cord. The relation between these kinds of intelligences appears to be that *bodily kinaesthetic intelligence, knowing how, underlies the development of all the rest.*

Indeed, it appears that the development of each of these kinds of intelligence, even the far reaches of abstract logical/mathematical intelligence, relies entirely upon the development of bodily kinaesthetic intelligence. Analyses of the cognitive operations evident in logical-mathematical learning, for example, show its beginnings in the unfolding of sensorimotor intelligence. A number of other relations between each of these kinds of intelligence have also been found, for example there is a clear relation between spatial intelligence and all the others. Spatial intelligence is temporally prior as well as a logically prior necessary condition to the other kinds of intelligence, centrally to logical-mathematical and bodily-kinaesthetic intelligence. A good mathematician will have a highly developed spatial intelligence as will a person with highly developed bodily-kinaesthetic intelligence.[35]

4.8. Bodily Kinaesthetic Intelligence

Knowing how is our bodily kinaesthetic intelligence (or just "bodily intelligence"), underlying all the other kinds of identified intelligence. Brief references to the other kinds may prove helpful to set forth and understand how bodily intelligence is related to them, and to obtain a useful classification of performances. To obtain a classification, we can look at *knowing how* in the following way: we can initially look at performances as objectively defined or characterized in terms of rules or prescriptions independently of anyone actually doing them, then we can analyze those performances from the point of view of the requirements of one who would know how to do them. In effect, I want to initially distinguish between the objective performances themselves and those who know how to do them, so as to determine the *epistemological* relations between them.

For example, one can ask whether or not there is only *one* way of doing a given performance or whether there are *many* ways. There are clearly many ways to communicate with others linguistically, exhibiting one's linguistic intelligence or know how. A potentially infinite number of different sentences can be formed from relatively few grammatical rules and a finite number of words. In logic and mathematics, there is usually more than one way to prove many if not most logical/mathematical theorems, and there may be many different starting points or positions in a proof. For example, one might be able to start a proof with a conditional, or (for the same theorem), one might be able to start with assuming the negation of the theorem, proceeding with a *reductio ad absurdum*. Of course, the end points or goals of the performances are to actually communicate with others and to end up with actually proving the theorem.

But though there may be many ways of doing some performances, there may be only one way of doing others. There may be only one way of balancing oneself on a tightrope, or taking aim with an M16-A2 at a precisely specified target and actually hitting that target. Where we rather uncritically define the term 'performance' as an intelligent way of doing something, while recognizing that there are many kinds of intelligence, we can tentatively classify intelligent performances into *single*-pathed and *multi*-pathed performances.

A path is a way of actually carrying out the doing. As implied, a path of a performance has a beginning and an end, a terminus. A single pathed performance is one that, once the initial point or position is chosen, there is only *one* route to the terminus; a multi-pathed performance is one that, once the initial point or position is chosen, there may be *many* routes to the terminus.

In rejecting the "intellectualist legend" inherited from Descartes, we can rule out that knowing how [to perform intelligently] is a tandem exercise of first considering rules, prescriptions, or propositions, then putting into practice what the rules, prescriptions, or propositions tell you to do. Even with performances where one might reasonably assume one must first know the rules before one can perform, such as the game of chess, for example, Ryle and the others have shown that it is possible for one to know how without knowing the rules in the sense of being able to explicitly state or formulate them. Moreover, their *knowing how* is not a mere matter of luck nor is it an instance of habit. That is, even where there is a clearly specifiable *knowledge that* or set of rules defining or characterizing quite specifically the step by step procedures for the actual doing of the performance, the argument is that one can still know how without knowing these rules in the sense of being able to either state or explicitly formulate them.

Clearly most of us become natural language speakers without first knowing the rules of our own languages. Indeed, we all come to know how to speak our own natural languages before we know how to read and write. Again, Ryle's point, as is also the point of others who argue along the same lines, is that there are *two* different kinds of intelligence involved in *knowledge that* and *knowing how.*

Moreover, the arguments and empirical evidence show that even where we do know the rules and can explicitly state or formulate them, it would still not necessarily follow that we would know how to do the task. That is, it would not necessarily follow from our *knowledge that* that we would know when, where, how, with whom, and in what appropriate way to *apply* the rules. We would not necessarily be able to do the task in a manner that showed we know how. This was Kant's point in his distinction between the *understanding* and *judgment.* Kant claimed that there is no rule governing "the power of rightly employing" the rules themselves, that such a power belongs *to the learner* himself.

In sum, knowing *that* rule or prescription is clearly not sufficient for knowing how to do those tasks. It is also neither necessary nor sufficient for knowing how to perform yet other kinds of tasks. When we become speakers of a natural language, knowing the rules is not a necessary condition for knowing how to speak it. Neither is it a sufficient condition. For example, even in learning our native language, but also in learning any second natural language, we become familiar with *inappropriate* use of our newly learned words, sentences, various kinds of expressions, intonations, and the rules governing them, revealing that we do not yet know how to speak that language. It is not that we do not yet know the rules and vocabulary of the language; it is that we do not yet know how to use them.

4.8.1. Knowing How Without Any Rules

These same arguments apply even to knowing how to do certain other kinds of performances which are more "artificial" or less a part of behavioral maturation than learning a native language. In his discussion of the game of chess, for example, Ryle makes it clear that *knowledge* how is exercised primarily in the actual moves that one makes or concedes and in the moves one avoids or vetoes. Of the boy who does not know the rules of chess but nonetheless knows how to play the game: "So long as he can observe the rules, we do not care if he cannot also formulate them. It is not what he does in his head or with his tongue, but what he does on the board that shows whether or not he knows the rules in the executive way of being able to apply them"[36]

Knowing how to play the game of chess, even if one cannot state or formulate the rules, is (among other things) knowing how to imagine and plan for possible alternative permitted moves and strategies on the part of one's opponent. Knowing how to observe the rules, apply them in the moves, is not given with one's *knowledge that* of the rules themselves. Indeed, this is impossible in the game of chess since there are in fact an immense number of possible moves, unlike a game such as tic, tac, toe, in which there are a very limited number of moves easily mastered and remembered. This applies equally well to most persons knowing how to speak their native language, as well as those already knowing the rules of the language. They can know how without knowing any rules which characterize their doing. Clearly, their *knowing how* is not a matter of chance or luck.

Moreover, such *knowing how*, though a matter of practice, is not a merely habitual action. When actions are done by habit they are done without heeding what one is doing, and habitual practices or performances are such that one tends to be a repetition of the others. Ryle's and others' emphasis upon manner of a performance as indicative of *knowing how* is intended to focus upon the intentional heeding by which something is done. It shows up in the timing and smoothness of the manner of the doing. It is by the manner of performance that we separate intelligent from non-intelligent performances.

Of course, natural or artificial languages and games such as chess are *prima facie* relatively clear examples of *rule-governed* activities or performances. Though there are rules which we can explicitly formulate which may define or characterize certain kinds of performances on certain levels, such as the game of chess and, presumably, being a speaker of any natural language, there are other kinds of intelligent performances for which this is not so.

We should be clear on what is being stated here: this is not to say that rules cannot be given to one preparatory to the actual doing of such an intelligent performance. It may even be the case that there are very general rules which can be given for one anticipating the doing of such a performance. However, it is explicitly formulated step by step procedures governing the *over-all entire actual doing* itself from beginning to end which cannot be given, though we may be able to describe, with some such performances, what a successful outcome of the doing might be.

For example, we cannot give explicitly formulated rules for balancing a pin on its head, though we might state that a successful outcome is keeping the pin balanced on its head. We cannot explicitly formulate step by step procedures for an over-all performance of balancing ourselves on a tightrope, nor can we explicitly formulate such rules for probing a wound without further injury to a victim, or for leading a platoon of Marines into any enemy position no matter how much we may know in advance of that position.

And if Kant is correct, there is no rule for the "power of rightly employing" rules themselves, even where we have them, such as with speaking one's natural language, playing a game of chess, or being kind to another human being. That cognitive power is found *in the person,* and is cultivated with task-oriented practices exhibiting bodily intelligence. It is in the *one* who knows how. It is *not* found in prescriptions, rules, or statements.

We must sort out performances which are explicitly characterized or defined by rules from those which are not, independently of persons who may perform them. Such a sorting may range across the kinds of intelligences noted above. But it is beyond my purposes here to provide an exhaustive classification applying to all six kinds of intelligence identified by Gardner, *et al*. We must also determine, from the side of the analysis of one who would perform, whether or not one's knowing those rules is either (1) necessary; (2) sufficient; (3) necessary and sufficient; or (4) neither necessary nor sufficient for the doing or knowing how to do the performance.

4.9. Classification of Performances

Thus we tentatively have other ways of classifying performances. Where "knowing that prescription or rule" means explicitly stating or formulating the step by step procedures of the rule, we have the following possibilities, intended to provide at least a working and useful classification:

(1) Performances for which knowing that (prescription or rule) is necessary but not sufficient for *knowing how*;
(2) Performances for which knowing that (prescription or rule) is sufficient but not necessary for *knowing how*;
(3) Performances for which knowing that (prescription or rule) is both necessary and sufficient for *knowing how*;
(4) Performances for which knowing that (prescription or rule) is neither necessary nor sufficient.

Examples falling under (1) might include performances found in the logical-mathematical kinds of intelligence. One must know rules for derivation, for example, to begin to prove theorems, though knowing the rules of derivation is not sufficient to know how to prove theorems or even to argue rationally,[37] (much to the very tired disappointment of those of us who teach logic to freshmen!) We may also include certain kinds of medical tasks here, where knowing the procedures for surgery may be necessary but they clearly are not sufficient for knowing how to perform surgery. Though it is clear that knowing the rules is not sufficient for knowing how to do any surgery, it is not even clear that knowing the rules in all cases is even necessary. There are known instances where persons not knowing the rules of procedure have performed such surgeries.

A point to keep in mind is simply that the intelligence involved in putting prescriptions or rules into practice is not identical with that intelligence involved in intellectually grasping or understanding the prescriptions or rules.

Examples falling under (2) might include linguistic performances. That is, being able to explicitly state or formulate a rule, at least in appropriate situations, we may claim is sufficient to show that one knows the language in which to so formulate the rule. But this is problematic because one could be taught to parrot or mimic the language to state the rule. The above discussion regarding knowing how to speak one's own natural language and learning a second natural language may suffice to show that even with knowing how to speak a natural language, one can know the rules yet not know how to apply them. Moreover, one may not know the rules at all, yet be a perfectly good speaker of a natural language. Much controversy is currently attached to the claim that a computer understands or knows how to speak certain languages because computers can simulate language rules to some degree.

If Kant, Ryle, Scheffler, Gardner, and others are correct, there are no performances falling under (3), though there are many falling under (4), including examples I have provided above. If we cross partition these four possibilities with single- and multi-pathed performances, we obtain the following matrix:

	KT nec for KH	KT suff for KH	KT nec+suff for KH	KT neither nec/suff For KH
Multi-pathed	?	0	0	√
Single-pathed	?	0	0	√

[Where 'KT' stands for *knowledge that* and 'KH' stands for *knowing how*.
'√' stands for "obtains"]

Figure FOUR-4. Classification of Performances

Boundary Set *S* consists of those kinds of *knowing how* for which *knowledge that* is neither necessary nor sufficient and which overlap with immediate awareness, *knowing the unique*. It includes performances which are both multi- and single-pathed and consist of those primitive immediate awareness relations discussed above embedded within the patterns of action of *knowing how*. As I have stated, *knowing how* is defined in terms of the actual manner with which one performs a task, with *manner* defined in terms of conditions of *smoothness* and timing of performance. Smoothness is further defined in terms of complex dynamic self-organizing sensory and somatosensory-motor oscillations which result in patterns of action with embedded primitive relations of immediate awareness, *knowing the unique*. I will return to a more precise mathematical discussion of these concepts below.

4.10. The Hierarchy of Primitive Relations of Immediate Awareness

The primitive structures of immediate awareness are a hierarchy (or stacked set of sheets) of primitive relations. When publicly manifested in an actual doing, they are within sign relations.[38] These primitive relations form the most basic variables or nodes of the hierarchy of a multilayer recurrent network of somatosensory-motor and sensory levels. Though the labels that I give to these elements are also sometimes used in ordinary language in a representational sense, I will try to make clear the distinctions between them.

The primitive structures are one class of the epistemic elements which for purposes here I will idealize later within a Boolean network model as simple binary variables. When these primitive elements of immediate awareness are coupled together in multilayer recurrent neural networks, we can use Boolean network theory to study the knowing behavior generated from these. That is we can trace *trajectories of knowing* as the system of epistemic elements of a given network state responds to combinations of signals from other elements.

The following primitive elements sorted below do not form a taxonomy, but are arranged in a hierarchy such that one relation is necessary to have before the others. Though primitive, we may understand their meanings through other terms which are defined. Just as the concepts *set* and *membership* are primitive and undefined, though we understand their meanings through concepts which are defined, so we can also understand the following primitive relations as well through other defined concepts.

4.11. Primitive Relations of *Preattending, Attending and the Problem with Paying Attention*

Earlier, we saw that there is striking evidence of cognitive activity during the preattentive phase in brain neural activity. I cited a number of experiments which clearly show evidence of intelligent behavior on the part of subjects even in the absence of awareness. The latter term is used by researchers in the sense of "awareness that," or "consciousness that," indicating later activity, minimally during the attention phase (and with language), usually accepted as the entry point in the "circle of cognition." The attention phase is usually demarcated as that point when normal human subjects are aware of their perceptions and can say (or otherwise indicate) that they are.

The preattentive phase, hence, is the most basic epistemic primitive relation between a subject and an object or configuration of objects. It is the most primitive, immediate form of awareness, in which we select an object from other objects. It is also clear that *in that immediate relation*, the selected objects are not members of classes. In the preattentive phase, subjects do not select or sort as classification machines would; we select *sui generis* objects, unlike any other.

Moreover, as certain of the studies I cited show, we may not know *that* we have done this selecting, and in fact our verbal reports about our awareness may very well correlate negatively with the success of our actual selecting, as for example in perceiving whether or not an animal is present in a masked scene. But studies on perception and language have tended to dominate neuropsychological and neurophysical research generally. Thus, kinaesthetic bodily intelligence is neglected as it pertains to the preattentive phase. As a consequence, with few exceptions, it is not entirely clear how our other senses and other primitive relations of immediate awareness are related to the preattentive phase, except that the sensory fields of the cortex analyze stimuli so that the sensory system can filter the stimuli and align the filtered sense qualities with the stimulus. That's what the preattentive phase does neurologically, but it is not entirely clear what it does cognitively with all the other senses. I conceive of the preattentive phase of primitive immediate awareness as obviously more fundamental, but also much broader in scope than the attention system.

In the attention system, comprised of both the activation system (reticulo-thalamo-cortical system) and attention, primitive relations may include those of "awareness that" selective attending, conscious sensation, but the lower level processes of the preattentive phase are still there. The activation system, which has an interesting history of all by itself, serves as a central activating system monitoring and regulating levels of excitation of the entire organism. It monitors and regulates itself as well as sensory and motor functions. Some have called it a sort of metasystem within the central nervous system CNS. [39] Neurons in the parietal, temporal, and frontal cortex, in addition to the region of the supplementary motor areas of field 6 (*e.g.* the frontal visual field), serve the attention system. Near these sensory fields is the sensory hand-arm field that also has attention functions, including aligning the body and sensory systems with the stimulus. It appears that the activation system (activation and attention) has control over an entire set of secondary sensomotoric fields for vision, hearing, and others, *distributed all over the cortex*, when it exercises its sensory-motor attention and coordination.

The primitive object selected in attention may be either an abstract or physical object of our experience. 'Experience' is clearly not limited to sensory experience as it also includes abstract objects of the mind such as images of non-existing things and mathematical objects. Sometimes the term 'attention' is used in a descriptive sense as an act of classification, but it should be apparent by now that that is not the sense I am referring to here. Primitive attending, with preattending, are prior logically necessary epistemic relations for all other primitive relations such as the hierarchically arranged, multi-layered relations of *sensing* [sight, touch, smell, hearing, tasting], *imagining*, *memory*, and the more complex primitive relations of *first-hand familiarity, involving moving, touching* and *recognizing*.

Paying Attention

I should make clear that the primitive relation of attending is not equivalent to the concept captured by the phrase "paying attention to." It seems to me that there are at least two senses of the root word 'attend' which are often confused in the literature and wholesale fallacies are committed by moving back and forth from one to the other because of uncritical assumptions about knowing and awareness generally.

Sometimes, 'attend' is taken solely in its "paying attention" sense. When Searle[40] states that we need to distinguish between those things that are at the center of our attention and those that are at the periphery, he narrows the concept to its "paying attention" sense, even while he implicitly recognizes the two senses with his examples. One example he gives is that of attending to a philosophical problem, which is clearly paying attention to a *knowledge that* object. However, when he references the feeling of the chair against his back and the tightness of his shoes to which he says he has not been [explicitly] attending, but which are nonetheless (he says) parts of his conscious awareness, then he implicitly recognizes the more primitive sense to which I am referring here. Moreover, one might ask, does his use of the phrase 'conscious awareness' reveal an implied recognition of (what else?) an *unconscious* awareness? Or is he also taking awareness only in its mediated sense of *knowledge that*, just as he does with *attending*, while also ignoring immediate awareness? Searle's uncritical use of 'attend', 'awareness', as well as 'consciousness' imply a deeper uncritical epistemology limited to *knowledge that*.

Though there may be some disadvantages to doing so,[41] I have chosen to keep the terminology found in the traditional philosophical and other literature on this topic. One finds my use of the primitive sense of attending as selecting an object out from all others as far back as Plato,[42] as well as in Descartes, and as recently as in Russell's work, and in contemporary neurophysiological literature.[43] Moreover, even *Webster's Encyclopedic Unabridged Dictionary*[44] makes reference to the primitive sense of selective attending which I am using, as a selective narrowing and focusing of consciousness and receptivity.

Thus, the primitive relation of *attending* is the selecting out of an object from all other objects. Moreover, the object is a particular, configuration of particulars, or, as I will tend to say, the individual selected. That individual, particular or configuration of particulars is unique, like no other, and that selecting does not imply a *reflection about* the object. For example, it does not imply reflecting *that* the object has a relation to the one selecting. *This selecting cannot be classificatory selecting because it does not depend upon invariant properties or attributes of the object selected as one among a class of such objects.* It depends upon cognizing particulars and relations among particulars of the object which are like no other regardless of the properties that object may have in common with other objects of any kind. This nonclassificatory primitive selecting has been noted in the neurophysiological research for some time[45] though it has been largely ignored until relatively recently, due to the increased interest in phenomenal consciousness and immediate awareness. I will address how this primitive attending proceeds in greater detail later, particularly in relation to other primitive relations of immediate awareness.

4.11.1. Indexicality: Primitive Sign Relations

It should also be noted that it is this primitive epistemic principle of *attending* which underlies all indexicality, a subject we cannot pursue in the depth it deserves here. It is sufficient for our purposes to recognize that knowers use indexes (indices) or *indicators,* that is *signs,* such as certain symbols, patterns of moving and touching, patterns of actions generally, and images to refer to or to disclose the *objects* of their knowing, either abstract or physical objects, or, more importantly, *to disclose the knowing itself.* Even if a knower is not a natural language speaker at all, that knower still uses physical or abstract indexes, such as images, moving and touching, to point to or disclose the physical or abstract objects of their knowing. They also use them to disclose their own knowing. It is clear that I am using the term 'sign' in a broader sense than the term 'symbol'. The category of all signs includes the category of (alphanumeric) symbol. This differs from some contemporary uses of the term, for example, Crick's.[46]

There is a fundamental and broader sense in which we more often *disclose rather than represent* [in declarative sentences or symbolic form] our knowing. This is why our know*ing* extends beyond our know*ledge*, and we need a comprehensive theory of indexicality extended beyond linguistic, symbolic indexicals to *non-linguistic* sign indexicals. Since linguistic indexes are primarily symbolic, an extension to non-linguistic indexicals will be an extension to signs such as physical gestures, the use of images, patterns of movement or touching, as well as kinds of intonation with the voice. This requires a prior theory of signs, a theory of the use of images and patterns of movements as indices.[47] I will later address how such an extension would affect a geometric view of the epistemic domain or universe, with subsequent changes in methods of inquiry.

4.11.2. Primitive Relations of the *Senses: Seeing, Feeling, Smelling, Tasting, Hearing* and *Imagining* to *Attending*

Before proceeding to the individual senses, I should clarify much more about the relation of attending. I believe this will help to see the relation of attending to the senses and to the spatial relation of each sense relative to the human body and relative to the object of knowing. We need to clarify the concepts *attending from* and *attending to*, as well as to other primitive elements. This will also help the reader to understand later the differences between what I refer to as the *rule-governed* nature of *knowledge that* and *rule-bound* nature of Boundary Set *S*, especially immediate awareness. To clarify the relation of attending itself, and its further relation to the individual senses, I will make use of the medical concept of *facies*, and will also refer to Polanyi's[48] notion of a *physiognomy*.

Physiognomy is a broad concept meant to apply to a broader range of epistemological phenomena than the pathological conditions to which *facies* often applies. For example, a physiognomy can refer to the multitude of particulars present in a unique configuration of a human face or body, such as the subtle, delicate array or configuration of particulars making up the unique features of a human face. In part, it is that configuration of particulars such that a given face is unlike any other. Indeed, face recognition (which is at a different hierarchical primitive level of immediate awareness), appears to involve both configurational and featural processing, involving a system that is separate from our visual processing system that identifies other physical objects such as trees, shoes, and tables.[49]

Physiognomy is used to refer to physical objects which we can recognize *but cannot describe* and cannot know uniquely from a set of propositions or descriptions. It can also refer to other objects of sense, such as the delicate, unique tastes of wines, teas, a particular orchestra and conductor's performance of Stravinsky's work, and so on, which only an expert can recognize. Moreover, it can refer to the *manner* of a performance, the refined, sensitive moving and perhaps touching, including timing, which make up *knowing how,* exhibited in the actual doing of a task or performance by one who knows how. One can see exhibitions of such knowing in dancing, figure skating, conducting an orchestra, and playing musical instruments sufficiently to play the works of great musical masters. The intelligence necessary to knowing the physiognomy of an object or set of objects requires a great deal of first-hand familiarity over time.

As noted above, *attending* is a primitive relation which functions to select or select an object from all else. Put differently, it is selecting a figure, a pattern or configuration of particulars, from its background [or, I will say, *ground*]. Again, attending to an object does not imply reflection about that object. We can *auditorially* attend to an object, a pattern of sounds, which we distinguish from noise which is the ground. This was demonstrated in the above experiment of Näätänen, *et al.* We can *gustatorially* attend to a taste from all others which may accompany an experience of eating or drinking; we also *visually* attend to an object or pattern of objects in an environment, and so on for the other senses.

But to attend *to* an object, configuration, or pattern of objects means, in part, that the particular(s) to which we attend *do not* correlate with those of the ground of that object. Those particulars of the ground must be sufficiently randomly distributed so that we distinguish the object, *which is a kind of order, a pattern,* from it. When those particulars of the ground are not sufficiently randomly distributed, our senses can be confused between the object and its ground. We may make this point clearer by noting that we can be confused auditorially, for example, when the sounds made by an insect mimic the sounds of its background, as some insects do mimic other sounds in the background so as to protect themselves. That is, the insects make sounds which themselves are sufficiently randomly distributed so as to fit into the background noise, the ground. Also, we can be confused visually by deliberate camouflage which breaks up the contours of patterns of an object so as to fit into the random distribution of particulars in a background. And there are like confusions for the other senses.[50]

Attending *to* an object with any of the senses singly or in combination, as in a focal awareness of an object, is possible only by virtue of our subsidiarially sensing its background *as* background. That is we attend *to* an object *from* an awareness of its ground. Essentially, the ground *acts as an index or pointer to the object.* This fact makes clear that our somatosensory capabilities have a *transactional relation* with whatever constitutes the context or environment of an object [particular] of the senses. Polanyi[51] has put the epistemic and transactional relation between figure and ground thusly:

> . . . whenever we are focusing our attention on a particular object, we are relying for doing so on our awareness of many things to which *we are not attending directly* at the moment, but which are functioning as compelling clues for the way the object of our attention will appear to our senses.

I prefer to use the term 'cue' for particulars in the immediate awareness relation. The significant meaning of 'cue' is that it is a *sign*, a feature or *signal* indicating (as an index) the something perceived. On the other hand, the term 'clue' is appropriate for a *representational* category, *i.e.* a description, in that it is a piece of evidence leading one to a solution of a problem. I believe that Polanyi conflated the representational and presentational epistemic categories. But his point helps to explain why we tend to overlook things that are unprecedented.[52] Without cues, without pointers or indexes, *we tend not to see them.*

Knowing a *physiognomy* is knowing a unique object, a particular or configuration of particulars unlike any other. It is knowing the object as *sui generis.* As stated above, preattending and attending are the most basic primitive relations to that object, prior necessary conditions to other primitive relations we may also have with that object. The latter include the relations of sense [tasting, smelling, feeling, seeing, hearing], imagining,[53] first-hand familiarity, which includes moving and touching, and recognition.[54] In reality, our actual primitive immediate awareness relation with an object usually consists of *a multifaceted, multi-layered set* of such primitive relations. One of my purposes here is to unfold that multifaceted set, revealing its structure. Knowing a unique physiognomy is at least to meet the following epistemic conditions: one must (i) recognize the particulars; and (ii) recognize the *relations among the particulars*. These conditions already assume, as prior necessary conditions, preattending and attending to the particular(s), and selecting the particular(s) from a ground. This may involve some or all of the senses and first-hand familiarity, involving moving and touching.

4.12. Multiple Spaces of Primitive Immediate Awareness

To unfold the structure of this knowing, however, we have to *epistemologically* sort the senses in terms of their *epistemic* spatial relation to the human body and also relative to an object of knowing. Recall that the posterior cortex contains multiple representations of space. This is neurological fact that has far-reaching epistemological significance. We may refer to those senses which are *close,* that is those which permit us to have an epistemic relation with objects near to the body, and those which are *distant*, that is those senses which allow us to have epistemic relations with objects which may be far away. Vision, for example, is a distance sense, while tasting and feeling are close.[55] With respect to distance, we may hear objects [sounds] which are far removed from our bodies, and we may also smell objects which are at a distance from us.

This spatial distinction between the senses has significance to natural intelligence in a variety of ways which we can only touch upon here. With respect to vision, for example, some objects or a unique configuration of a thing [set of particulars] may be seen without effort at a distance, from a more "global" perspective, but cannot be seen at all locally or close up. That is, the particulars close at hand do not form a *visually* recognizable pattern or configuration because we do not see the relations among the particulars. A clear example of this is the patterns of objects on the Nazca plain in Peru. These were not recognized at all from the ground, but are seen only at a higher elevation and greater distance from the particulars forming the patterns. In this sense, we may say that the object, that is configuration of particulars, is recognizable when seen at a distance only because the relations among the particulars is recognizable at a distance and not close up. This concept of immediate awareness may overlap with an example of recognition in its representational sense, if we are able to represent the relations in some symbolic form. For example, we can mathematically characterize the configuration of objects on the Nazca plain as geometric forms.

However, even when the relations among the particulars are seen at a distance, and a form recognized, there is still an *unspecifiability* of the particulars themselves both close up and as well at a distance. It is *the relations* among the particulars forming the patterns which are describable, once seen. But close up, we may only be aware of particulars, not the relations among them. Uncountable numbers of those particulars remain as the ground of the object, the form or pattern (configuration) of relations among particulars, which is seen only at a distance. If sufficiently randomly distributed, and if the person is sufficiently at a distance, these particulars of the ground function indexically to point to the object, *they make possible the "signature" of the object.*

This sense of spatial distance enabling the awareness of a configuration of an object underlies the mathematical notion of *manifold*. Manifolds are objects studied in an area of mathematics called *topology*, which studies the properties of objects when they are changed. They may be changed when twisted or stretched, or made larger or smaller. Topology tries to understand both the local and "global" properties of these objects, and whether or not two objects that may look very different on one scale are in fact the same from a mathematical perspective on a different scale.

On small scales, an object may look to be one thing, while on a more "global" scale it may turn out to be entirely different. A good example is the earth itself, which locally looks "flat," but is in fact round (more or less). We say that it is "locally Euclidean" because it has properties describable with the concepts and tools of Euclidean geometry. Euclidean objects like circles and spheres are manifolds, and topologists have determined that a circle is topologically equivalent to an ellipse. This is so because we can stretch the circle into an ellipse. The space of all positions on a minute hand of a clock is topologically equivalent to a circle, while the space of all positions of both the minute and hour hands is equivalent to a torus (like a donut). But a significant point to be made here is that topology treats spatial objects such as circles and spheres as objects in their own right. Our knowledge of objects is *independent of how they are represented* or embedded in space.[56]

The epistemological significance of the spatial relations our bodies have with an object might also be illustrated by the highly complex task of viewing tissue cells under a microscope. Sometimes the physiognomy [and sometimes even the identity] of the cells of a tissue specimen cannot be recognized if the cells are viewed *at certain angles* under a microscope. However, once one turns the slide a bit to one angle or another, the physiognomy or identity can be immediately recognized. That is, when we change the spatial arrangement of the object in relation to ourselves, we can sometimes recognize the object. Again, distance and spatial configuration in relationship to our bodies play an epistemological part in our knowing of an object.

These examples point to the fact that an *analysis of particulars*, taken separately in isolation or as atomic units, cannot enable us to cognitively or epistemically "capture," perceive, or know the comprehensive entity of which they are a part. That is, we cannot come to know an object solely by means of analysis of that object's particulars. This continues to be an underlying assumption of the atomistic, summative, or what I call the "building blocks" view of knowing, and the stimulus-response view of natural intelligence. An analysis of an object's particulars will not lead to an understanding of the relations among the particulars and moreover, the *emergent properties* of the relations among the particulars.[57] We will not know them, thus we will not know the object. Both analysis and synthesis or integration of the particulars in terms of their relations are required. But there will always be a residue or multiplicity of particulars which escape analysis altogether, that cannot be specified, defined, or described, because they form the ground of the object. Nonetheless, they play a fundamental epistemic transactional, indexical role in our knowing of these objects as unique.

4.13. The Primitive Relation of *Imagining*; Hierarchy of the *Senses, Touching, Moving*, Probing and their Spaces

The senses, imagining, touching and moving form a multi-relational, multi-layered hierarchy, with touching, imagining and moving on a higher level than any of the other senses, except sight. These are represented in the following graph. Though I have included recognition, I will not thoroughly discuss it in this work.

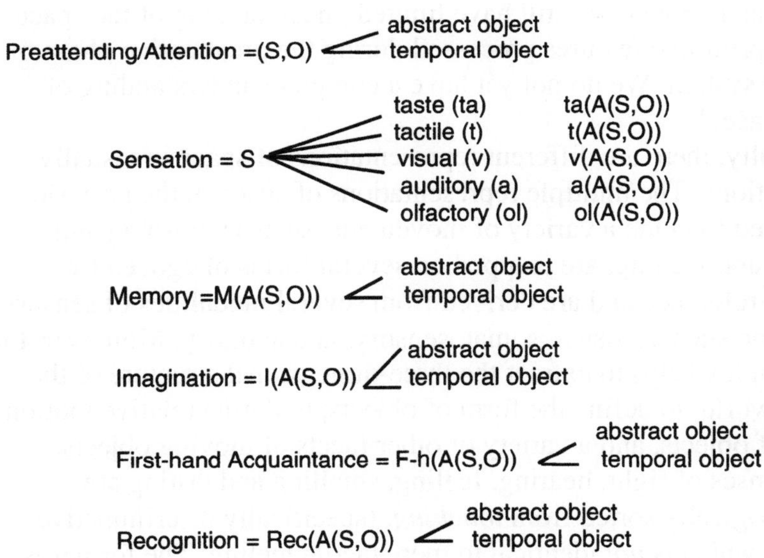

Figure FOUR-5. Multi-layered Hierarchy of Primitive Relations of Immediate Awareness

A clarification of these should help to understand a little better the significance of the nature of *probing*. However, though I focus here upon physical probing in a medical surgical task, I hope it is clear that I do not limit probing in general to physical spaces and with sense [physical] objects. We also probe abstract spaces, as mathematicians and logicians clearly evidence. Both Russell and Gödel, for example, have pointed out the close analogy between how we know reality with our senses and how we know abstract objects of mathematics with our minds. They both recognized the significance of immediate awareness in mathematical knowing.

With respect to the particular senses, seeing, hearing, feeling, tasting, smelling, there are a number of principles pertaining to these which require explanation over and above the spatial relation each sense has with our bodies. Firstly, not all of the senses are on the same primitive *epistemic* hierarchical level. In physiologically unimpaired persons, the sense of sight takes some priority over the other senses, and it is clear that the space of our visual experience is not identical to the space of the other senses. For example, visual space is *binocular* space, while the space of the other senses, for example smell, is not. But as already noted, we still have limited understanding of the space of all the primitive features processed during the preattentive phase of the visual system. We do not yet have a complete understanding of "shape space."

Additionally, there are different representations of space in visually guided actions. The multiple representations of space in the posterior cortex, used to guide a variety of movements such as grasping and reaching, and feeding, are mapped on several forms of egocentric frames of reference and are derived from several modalities of sensory information such as visual, somatosensory, and auditory. Moreover, the MT+ complex helps to extract the three-dimensional structure of the physical world, to define the form of objects, to define relative motion of parts of objects, and a variety of other facets of moving objects.

The senses of sight, hearing, feeling, smelling and tasting are epistemologically sorted from *touching*, (specifically discriminative touching) which is *not* identical to mere tactile feeling. The former is clearly intentional while the latter is not. Eccles cited an excellent experiment effectively showing the difference between the two. The experiment showed the effect of silent thinking on the cerebral cortex, in which a subject was "concentratedly attending to a finger on which just detectable touch stimuli were to be applied. There was an increase in the rCBF [rate of cerebral blood flow] over the finger touch area of the postcentral gyrus of the cerebral cortex. These increases must have resulted from purely *mental* attention because actually no touch was applied. . ."[58]

These senses are also sorted from *moving*, which is treated as a complex sensory *and* somatosensory-motor phenomenon in the neurophysiological literature,[59] involving the above different representations of space in visually guided actions. I include moving with touching as primitives at a level higher and distinct from though including the other senses, including tactile feeling. In part, this is because the concept *touching* is clearly *bodily* intentional, in the sense that we use our bodies cognitively to index in kinds of space when we touch. Whereas mere tactile feeling is not used this way.[60] Moreover, the space of feeling and the space of touching are not identical. For example, intentional touching is not clearly always in Euclidean three-dimensional space because of the relation of imagining, including *anticipatory* imagining to it, as we will see below. However, mere tactile feeling [such as feeling a pin prick] clearly is in Euclidean space. A pin prick is felt *here*, *now*, *in this space that I can physically point to.*[61]

The senses and the concepts *touching* and *moving* are enormously complex concepts which will be evident in our analysis below of knowing how to probe in a surgical task. One way of distinguishing the senses from touching and moving and other primitive epistemic relations, including imagining, is to note that the objects of the senses, that is particular sights, smells, tastes, or configurations of these, are exactly that. That is they are particulars which occur "now" with the subject. They are not universals or generalizations publicly accessible to anyone, though in principle the same particular may be experienced by more than one person.[62]

The intentional concept of *touching* is more complex in that it entails a deliberateness with the body which is not found with the senses *per se,*[63] and where imagining is involved there may be abstract universals which may be experienced by more than one person. A thorough analysis of the epistemic structure of *touching* requires an analysis of probes and their epistemic relation to our body. Moreover, what we know of the human use of the fingers to explore or come to know the texture and shape of objects has much in common with results of scientific neural experimentation with the rat trigeminal system. We know that rats rely on rhythmic movements of their facial whiskers *much as humans rely on coordinated movements of fingertips* to explore or come to know objects in their proximal environment. The trigeminal system is a multilevel, recurrently interconnected neural network which generates complex emergent dynamic patterns of neural activity manifesting synchronous oscillations and even chaotic behavior,[64] as found in Boundary Set *S* generally.

There is an epistemological sense in which touching requires that one intentionally *heed* and *focus* upon the object of touching with one's body, whereas one can experience with one's senses without that kind of intentionality. Moreover, this intentional heeding and focusing will differ epistemically in its structure depending upon whether or not one has unaided visual access to the object(s) to which one is heeding or focusing. It will also differ depending upon whether or not one is touching the object directly with one's body [for example with a hand or finger], or if the touching is mediated by an instrument used as a mechanical probe of some kind.

For example, if one is probing with one's fingers the interstices of a surgical incision not visually accessible, the epistemic structure of that *coming to know,* the probing, differs from the digital inspection of a wound which is visually accessible. This is so in part because the structure of the former requires more complex relations of imagining. As we will see, such probing of the inside of a wound requires *continuously forming images of the object*, that is the particulars making up the configuration of the inside of the wound, as touching proceeds.

The reference to 'as touching proceeds' indicates the complex nature of the relation of the epistemic structure of *moving* with one's body [or a part of it] to the already complex epistemic structure of *touching*. Moving which has epistemological significance is clearly intentional and requires a focal heeding with one's body, as does touching. However, it is not clear whether the space of moving is equivalent to or identical with the space of touching because of the relation of imagining [and especially imagining which is anticipatory] to the latter, and also because touching is more a *close*[65] relation than moving which can be more distant. With touching, the body is clearly used indexically in a very close, concrete way with proximal objects. With moving, the indexical function of the body may be more abstract because it can involve objects (including imagined and anticipated patterns) which are at great distances from the body.

The same kind of heeding or focusing found with touching and moving is not found with experiences of the other senses precisely because of the unique digital or indexical use of the body, and the spatial relations of touching and moving with the body. Moreover, though touching may involve any part of our body, the fingers *as digital indexes* are pivotally involved in an epistemic sense, as a means of directing our *coming to know* an object of touch. As already noted, there is an epistemic sense in which we extend our body to include the object presented, to which we attend, *from* the [sometimes imagined and anticipated] focal and subsidiary configurations of particulars we are aware of with our fingers. And as our analysis of probing an open wound with one's fingers will show, the relation of imagining is also pivotally involved in touching, in tactile efforts to come to know the object of touch. Not only are images formed of configurations of physical particulars presented, but images are also formed of abstract configurations of particulars *anticipated.*

For these reasons, as referenced above and elaborated upon, I have distinguished between *rule-governed* knowing, which is *knowledge that*, definable in terms of explicit natural language or logic-based rules by which we come to know *that;* and *rule-bound* knowing which transcends such explicit rules altogether. Nonetheless, *rule-boundedness* of such knowing may be mathematically characterized or simulated to some degree with recurrent self-organizing multilayer neural network models of knowing behavior, generated by distributed and dynamic parallel-processing mappings.

4.14. Summary

This chapter presented evidence to show that there is a level of immediate awareness that *is* cognitive and *is not* mediated by statements, symbols, or linguistic units of any kind. In fact, evidence from several studies show that this level of awareness correlates negatively with verbal reports or encoding. That level of awareness is found minimally in the preattentive phase of neural activity in the central nervous system and overlaps with attention. I argued against those interpretations of all neural activity as symbolic or "representational," due to their confusion between symbols and the things symbolized, or between "representations" and the things "represented." Such confusion results in collapsing levels of analysis and begging questions at issue. The primitive relations between subject and object at the level of immediate awareness cannot be one between a subject and a class object because the object(s) *in that primitive relation* are not members of classes. They are *sui generis* objects in that primitive relation with the subject.

The general concept awareness is not viewed in terms of two mutually exclusive states, awareness or unawareness, but is viewed as a continuum of states ranging from unaware through an infinite number of partially aware states, to complete awareness. This continuum, however, also distinguishes between "awareness that" such and such is the case [tying awareness to "that" clauses or linguistic reports] and "immediate" awareness which is not tied to such reports. What is needed is a clear map of where the two categories of awareness lay on the continuum. Sorting a hierarchy of primitive relations of awareness, including those of immediate awareness, showing where they lay on the continuum may provide us with that map. This is an issue to be addressed in greater detail later.

Arguments were also presented raising issues with the description of the preattentive phase and with the use of the concepts 'conscious' and 'attention', and distinctions between cognitive and non-cognitive. Because the organism is already making *preparations* and *aligning its senses* with some stimulus during the preattentive phase, this logically implies that the organism is *already directing itself* in ways to attend to some stimulus that it has already *in some more primitive sense* selected to align itself with. It has to have made such a selection since any given stimulus would be in an environment filled with possibly an infinite number of stimuli from which to select. The preattentive phase is said to precede conscious sensation in the activation of attention, with the use of the term 'conscious' tied to "awareness that" such and such is the case. And it is in the attention system combined with the activation system that, so it is claimed, cognition occurs.

I cited a number of experiments, however, that reveal, among other things, that we must revise our understanding of the cognitive domain and the place where we enter it. Currently, cognition is viewed as largely starting with the attention system and continuing on to "higher" levels, that are one way or another aligned with language. But all of the experiments cited showed some deeper level of awareness, below the attention system threshold, that *correctly* affected subjects' overall behavioral responses. As noted, some of the experiments also show that there is in fact a *negative correlation* between subjects' own verbal judgement (*knowledge that*) about their own awareness and their awareness as actually measured in the experiments. This evidence shows that the circle of cognition is larger and deeper than previously thought. This is so as it pertains to not only vision, but also the psychomotor and entire sensory motor parts of the brain. Thus I argued that not only do we need to revise our understanding of the scope and depth of the cognitive domain and the place where we enter it, we must also revise our understanding of a network of related concepts, including cognition itself, natural intelligence, learning, and conditioning. If the empirical findings and our interpretations of them are correct, natural intelligence begins with immediate awareness in the preattentive phase.

On the basis of relatively recent experiments involving the spinal cord after injury, I also argued that natural intelligence involves not just the brain but the entire central nervous system. Arguing that highly complex, dynamic interactions among primitive relations of the entire sensory and somatosensory-motor systems are involved in natural intelligence behavior, I cited findings of Gardner who identified at least six separate and distinct kinds of natural intelligence, basing his research primarily upon neurological, cross-cultural, and psychometric evidence. These kinds of intelligence include *linguistic, musical, logical-mathematical, spatial, bodily-kinaesthetic*, and *personal* intelligences, involving different parts (sometimes overlapping) of the central nervous system, both brain and spinal cord. The relation between these kinds of intelligences appears to be that *bodily kinaesthetic intelligence, knowing how,* underlies the development of all the rest.

To better understand the way bodily kinaesthetic intelligence, knowing how, underlies the development of all natural intelligence, I presented a working classification of performances. Initially, sorting them as either multi- or single-pathed, I later sorted performances into groups which are explicitly characterized or defined by rules from those which are not, independently of persons who may perform them.

For our purposes, I wanted to determine, from the side of the analysis of one who would perform, whether or not one's knowing those rules is either (1) necessary; (2) sufficient; (3) necessary and sufficient; or (4) neither necessary nor sufficient, for the doing or knowing how to do the performance.

I also set forth a tentative definition of Boundary Set *S* in terms of that classification. Boundary Set *S* consists of those kinds of *knowing how* for which *knowledge that* is neither necessary nor sufficient and which overlap with immediate awareness, *knowing the unique*. It includes performances which are both multi- and single-pathed and consist of those primitive immediate awareness relations discussed above embedded within patterns of action of *knowing how*.

A hierarchy of primitive relations of immediate awareness were also sorted, particularly in terms of multiple kinds of space. They were arranged in a hierarchy such that one relation is necessary to have before the others. Both the classification of performances and hierarchical classification of the primitive relations of immediate awareness are in preparation for a much more formal treatment of their highly complex and dynamic interrelations to be presented in the following chapters. Bodily kinaesthetic intelligence, knowing how, was analyzed in terms of kinds of performances and in terms of some of the other senses, especially moving and touching, though a thorough analysis of the epistemic structure of *touching* requires an analysis of probes and their spatial relation to our body and our use of images.

More analysis of the complex, dynamic relations between these categories of primitive immediate awareness and knowing how will continue in the next chapter, to better place this analysis in a theoretical and mathematical framework.

---

[1] Kurt Gödel, "What is Cantor's Continuum Problem?" in *Philosophy of Mathematics, Selected Readings*, Paul Benacerraf and Hilary Putnam (eds.), Prentice-Hall, Inc., 1964.

[2] M. Bjorkman, P. Juslin, & A. Winman, "Realism of confidence in sensory discrimination: The underconfidence Phenomenon," in *Perception & Psychophysics*, Vol. 54, 1993, pp. 75-81. This discussion of definitions of awareness is based largely on Kunimoto, C., *et al.*, "Confidence and Accuracy in Near-Threshold Discrimination Responses," in *Consciousness and Cognition*, Vol. 10, no. 3, 2001, pp. 294-340.

[3] C. Kunimoto, *et al.*, "Confidence and Accuracy in Near-Threshold Discrimination Responses," in *Consciousness and Cognition*, Vol. 10, no. 3, 2001, pp. 303-304.

[4] C. Kunimoto, *et al.*, 2001, p. 296.

[5] B. Libet, "Electrical Stimulation of Cortex in Human Subjects, and Conscious Memory Aspects," in A. Iggo (eds.), *Handbook of Sensory Physiology*, Vol. II, Springer-Verlag, Berlin, Heidelberg, New York, 1973, pp. 743-790.

[6] In much of the neurological literature, the terms 'aware' and 'conscious' are often used to describe a subject who is awake and can respond, especially in language.

[7] An excellent account of these can be found in Karl R. Popper and John C. Eccles, *The Self and Its Brain*, Springer-International, 1977, pp. 256-259 f.

[8] Karl R. Popper and John C. Eccles, *The Self and Its Brain*, 1977, p. 362.

[9] R. Hernegger, "Changes of Paradigm in Consciousness Research: Phylogenesis of Symbolic Information," 1995, emphasis mine.

[10] R. Hernegger, "Changes of Paradigm in Consciousness Research," 1995. I have relied on Hernegger for much of this section.

[11] J. Wolfe, "Visual Search," in H. Pashler (ed.), Attention, London: University College of London Press, 1996.

[12] J.M. Wolfe, "Preattentive Object Files: Shapeless Bundles of Basic Features," in Vision Research, Volume 37, Issue 1, January, 1997.

[13] J. M. Wolfe, "Visual Search," in H. Pashler (ed.), *Attention*, London: University College of London Press. [Note his use of the word 'know'].

[14] There may be exceptions to this, as for example in texture search. See Christopher Healey, *Pre-Attentive Processing*, University Of North Carolina, 1993.

[15] Ibid.

[16] See Wolfe, Ibid.

[17] J.M. Wolfe, "Visual Search," in H. Pashler (ed.), *Attention*, London: University College of London Press, 1996.

[18] Fei Fei Li, Rufin Van Rullen, Christof Koch & Pietro Perona, "Rapid natural scene categorization in the near absence of Awareness" [Also: "Rapid Visual Categorization in the Absence of Awareness," in *Proc Nat Acad Sci,* vol 99, July 2002]; also: John Colombo, Jennifer S. Ryther, Janet Frick, Jennifer Gifford, "Visual Pop-out in Infants: Evidence for Preattentive Search in 3- and 4-month-olds," *Psychonomic Bulletin & Review*, Vol 2, number 2, June, 1995, pp. 266-268; Risto Näätänen, Mari Tervaniemi, Elyse Sussman, Petri Paavilinen and Istvan Winkler, "'Primitive Intelligence' in the Auditory Cortex," *Trends in Neurosciences*, Vol 24, number 5, 2001, pp. 283-288.

[19] John Colombo, Jennifer S. Ryther, Janet Frick, Jennifer Gifford, "Visual Pop-out in Infants: Evidence for Preattentive Search in 3- and 4-month-olds," *Psychonomic Bulletin & Review*, Vol 2, number 2, June, 1995, pp. 266-268.

[20] Craig Kunimoto, Jeff Miller, Harold Pashler, "Confidence and Accuracy in Near-Threshold Responses," in *Consciousness and Cognition*, Vol. 10, number 3, pp. 294-340.

[21] Repp, Bruno, "Phase Correction, Phase Resetting, and Phase Shifts After Subliminal Timing Perturbations in Sensorimotor Synchronization," *Journal of Experimental Psychology: Human Perception and Performance*, APA, Vol. 27, Number 3, June, 2001.

[22] John Colombo, Jennifer S. Ryther, Janet Frick, Jennifer Gifford, "Visual Pop-out in Infants: Evidence for Preattentive Search in 3- and 4-month-olds," *Psychonomic Bulletin & Review*, Vol 2, number 2, June, 1995, pp. 266-268

[23] Risto Näätänen, Mari Tervaniemi, Elyse Sussman, Petri Paavilinen and Istvan Winkler, "'Primitive Intelligence' in the Auditory Cortex," *Trends in Neurosciences*, Vol 24, number 5, 2001, pp. 283-288

[24] See Lawrence Weiskrantz, *Consciousness Lost and Found*, Oxford University Press, 1997.

[25] These experiments are in part referenced in Michael Polanyi, *The Tacit Dimension*, Doubleday & Company, Inc., 1966. More recently, Gavin de Becker has documented an enormous amount of information related to how we accurately become aware of the

presence of danger without relying upon knowledge or awareness *that* (He does not use the phrase "knowledge that."). See his *The Gift of Fear*, New York, Dell Publishing, 1997.

[26] M. Livingstone and D. Hubel, "Segregation of Form, Color, Movement, and Depth: Anatomy, Physiology, and Perception," in *Science*, Vol. 240, 1988, pp. 740-749.

[27] Tutis Vilis, *The Physiology of the Senses: Transformations for Perception and Action*, University of Western Ontario, 2002.

[28] Peter W. Jusczyk, *The Discovery of Spoken Language*, Cambridge: MIT Press, 1997.

[29] Jody C. Culham and Nancy G. Kanwisher, "Neuroimaging of Cognitive Functions in Human Parietal Cortex, in *Current Opinion in Neurobiology*, Vol 11, 2001, pp. 157-163.

[30]Luba Vikhanski, *In Search of the Lost Cord*, Joseph Henry Press, Washington, D.C., 2001. My discussion of the spinal cord is largely based upon Vikhanski's work.

[31] Luba Vikhanski, 2001, pp. 180-185. Conditioning is usually defined in terms of behavior modification. A subject comes to associate a behavior with a previously unrelated stimulus. Thus, there is a cause-effect association established in the brain after repeated trials. See *The American Heritage College Dictionary*, Third Edition, Houghton Mifflin Company, 1993.

[32] Jonathan R. Wolpaw, "The Complex Structure of a Simple Memory," in *Trends in Neurosciences*, Vol. 20, 1997, pp. 588-594.

[33] Luba Vikhanski, 2001, p. 185.

[34] Howard Gardner, *Frames of Mind: The Theory of Multiple Intelligences*, Basic Books, 1993. Also see his 1973,1978,1982,1983, 1985.

[35] Among the six kinds of intelligences, linguistic and logical-mathematical have undoubtedly been studied more than the others. There are still disputes concerning how we learn or come to know language, but it is evident that within a very few years following birth, most normal children will be able to engage rather well in their natural languages. I will have little to say about this other than arguments I already presented in the discussion of Quine's theory, given that my concerns are with kinds of knowing found in knowing how and immediate awareness.

[36]Ryle, 1949, p. 41.

[37]See Ryle, 1949, pp. 47-50.

[38]The term 'presentation' is uniquely appropriate in many ways which will become clear later as I analyze a medical task found in the intersection of knowing how and knowing the unique. Though not limited to the practice of medicine, medical practitioners speak of patients as *presenting* with certain *signs* and symptoms of disease or other pathological condition. As the concept *facies* shows, those presented signs are often not reducible to representations (or descriptions).

[39] R. Hernegger, R. "Change of Paradigms in Consciousness Research," 1995.

[40]John Searle, 1992, pp. 137-138.

[41]The disadvantage comes from the tendency of some to do *ad hoc* disengagements of the traditional meanings of terms from those traditions.

[42]See "Theaetetus" in *The Philosophy of Plato*, The Jowett Translation, Irwin Edman (ed.), New York, The Modern Library, 1928. The discussion between Theaetetus and Socrates in which Socrates uses the image of a wax impression entails the primitive sense to which I am referring.

[43]See "The Organization of Perceptual Systems," in *Perception: Mechanisms and Models*, Readings from Scientific American, San Francisco, W.H. Freeman and Company, 1972.

[44]*Webster's Encyclopedic Unabridged Dictionary*, New York, Portland House, 1989.

[45]For example, one finds extensive consideration of this in Cherry, 1957.

[46]Francis Crick, *The Astonishing Hypothesis*, New York, Simon & Schuster, 1995.

[47]Indeed, recent empirical evidence [see Tanenhaus, *et al.*, "Integration of Visual and Linguistic Information in Spoken Language Comprehension," in *Science*, Vol. 268, 16 June 1995, pp.

1632-1634.] shows that nonlinguistic visual imagery affects the manner in which linguistic input is initially structured and comprehended. Ikons appear to be necessary to symbols in language comprehension.

[48]See Michael Polanyi, *Knowing and Being*, Marjorie Grene, (ed.), The University of Chicago Press, 1969.

[49] See Stephen M. Collinshow and Graham J. Hole, "Featural and configurational processes in the recognition of faces of different familiarity," in *Perception 2000*, Volume 29, number 8, pp. 893-909.

[50] Polanyi, 1969, Ibid.

[51]Polanyi, Ibid., p. 113. He uses the term 'clue' while I prefer to use the term 'cue' for particulars in the immediate awareness relation.

[52]This is a point also made by Ludwig Wittgenstein in his *Philosophical Investigations*, p. 50, #129. Obviously, arguing that a background functions indexically relative to a figure implies that noise actually serves a useful purpose even though it is normally considered an engineering nuisance. This same point is raised elsewhere in studies of the visual cortex [see "Separating Figure from Ground with a Bolzmann Machine" by Terrence J. Sejnowski and Geoffrey E. Hinton in Michael Arbib and Allen Hanson (eds.), *Vision, Brain, and Cooperative Computation*, MIT Press, 1987, 1990], and in more recent studies of noise in biological sensing systems [see "The Benefits of Background Noise," by Frank Moss and Kurt Wiesenfeld, in *Scientific American*, August, 1995, pp. 66-69]. Far more research needs to be conducted on the indexical use of signs in knowing systems, both natural and artificial.

[53]I use the term 'imagining' here rather than 'imaging' because the latter is too tied to *representation* in symbols. By 'imagining' I mean the cognitive or epistemic principle of forming an image "in the mind" not present to the senses or never before wholly perceived in reality. Imagining is a primitive epistemic relation with an image. That relation may be non-temporal, with abstract objects, in the sense that the image formed may not be in time at all, *e.g.*, mathematical objects, or a possible configuration of a wound which presents *with signs and symptoms*. Obviously, imagining may be a kind of seeing which cannot be accounted for by neurophysico-chemical and sensory accounts of visual processing.

[54]I have introduced the concept *recognition*, which has meaning in a relation of presentation (and also another sense of recognition which has meaning in a relation of representation), but I defer an analysis of it for a later publication. For now, we can understand recognition in its presentational sense [immediate awareness] as knowing a set of particulars unique to an object.

[55]Specifically, this is a distinction between the somatosensory system and the senses in general, but I am focusing upon its epistemological significance.

[56] Todd Rowland, "Manifold," *Eric Weisstein's Math World*, Wolfram Research, Inc., 1999-2002.

[57]This example serves also to demonstrate the inadequacies of sense datum approaches to epistemological inquiry.

[58] John Eccles, "The Effect of Silent Thinking on the Cerebral Cortex," in *Truth Journal*, Leadership U., 2002, p. 2.

[59]See the reference to Berthoz and Israël.

[60]Substantial empirical research has established this claim, in addition to that of Berthoz and Israël. See Gardner's *Frames of Mind: The Theory of Multiple Intelligences,* Basic Books, 1993. See especially references included under bodily-kinaesthetic intelligence.

[61]Because of the diversity and complexity of kinds of space characterizing the primitive features in the preattentive phase, the other senses, as well as touching and moving, I have chosen to limit the discussion here. A full treatment of these spaces would require a separate book.

[62] That is, these particular objects of the senses necessarily have a temporal relation with the subject who is having particular sensations, but only in principle can two subjects experience the same particular object of the senses, such as a particular color or taste.

[63]My efforts toward an analysis of the concept *moving* here, especially in relation to *touching*, cannot be complete. I have focused upon the concept insofar as it is epistemically involved in our task of probing an open wound. It should be noted that our scientific knowledge and understanding of human moving [movement generally, or whole-body displacement] is quite limited. We do not as yet even understand how moving is stored in the memory, how we spatially image or reconstruct a trajectory path [path integration] in our minds, or how we "home" in on a target, objective or goal with our bodily movements. See Alain Berthoz, Isabelle Israël, *et al*, "Spatial Memory of Body Linear Displacement: What is Being Stored?" in *Science*, AAAS, Vol. 269, 7 July 1995, pp. 95-98.

[64]Nicolelis, *et al.*, "Sensorimotor Encoding by Synchronous Neural Ensemble Activity at Multiple Levels of the Somatosensory System," in *Science,* AAAS, Vol. 268, 2 June 1995.

[65]Again, the terms 'close' and 'distant' as related to epistemic relations have meaning in relation to proximity with the human body, the ultimate instrument of all our external knowing. I am not happy with the distinction between touching and moving as I have left it here, and am not resigned to the distinctions between them as I have drawn them.

5. BOUNDARY SET *S*: AT THE CORE OF MULTIPLE INTELLIGENCES

In this chapter, I will provide a precise definition of Boundary Set *S*, setting forth the theoretical and mathematical framework within which to approach immediate awareness and knowing how. For those who find mathematical notation or formulas burdensome, please feel free to simply skip over to sections with which you feel more comfortable. I have tried to get at the more "intuitive" ideas underlying the formal parts by using some illustrations or analogies so as to make the following chapters more reader-friendly.

In the last chapter, we sorted out a working classification of performances in which *knowing how* with embedded immediate awareness, is exhibited. Using that classification, I will sort out and set aside conditions of *knowledge that* from those bodily kinaesthetic performances of *knowing how* and will analyze some actual performances showing the hierarchy of primitive relations of immediate awareness. The analysis will primarily be directed to the primitive relations of *touching* and *moving* of the somatosensory-motor system, embedded within those knowing how performances. We sorted those to some degree in the previous chapter, but will seek to refine that classification and hierarchy here.

Though I have made extensive references above to several kinds of performances and elaborated upon the primitive relations found in them, we do not yet have a formal view of the relations obtaining between these and kinds of *knowing how*. It is that formal framework I want to establish in this chapter.

5.1. Kinds of Knowing in Boundary Set *S*

Kinds of knowing found in Boundary Set *S* show up in the smoothly timed patterns of moving and touching by one who knows how. These smoothly timed patterns emerge from the complex dynamics involved in interactions of very large numbers of components and relations between them. This can be concretely demonstrated by analyzing the interactions among those components of very complex tasks.

For example, we can focus upon an earlier mentioned multi-pathed task, a surgeon softly probing an open wound for an unseen jagged projectile [shrapnel] without further injury to a victim.[1] It could just as easily be the task of probing an open incision for a diseased appendix, or any number of other such tasks performed everyday by medical practitioners. Any surgical task is very complex and one's knowing how to perform it is a complex, dynamic, and self-organizing kind of knowing. It is a motion-filled phenomenon exhibiting a very high degree of complexity resulting from very large numbers of diverse kinds of elements that are intricately interacting in very complex ways. When performed correctly by a person with appropriate medical expertise and accumulated first-hand experience and practice over time, such tasks exhibit highly complex, emergent properties that we see in smoothly timed performances.

These complex, emergent properties include knowing how to respond to unexpected events both internally to the actual performances themselves, for example the occurrence of respiratory failure, and in the surrounding context or environment in which the performance must take place. Medical practitioners performing such tasks in an antiseptic environment such as a hospital behave differently from those performing the same task (but with different procedures and equipment) in a combat environment[2] where extreme conditions sometimes require extensive innovation in ways of saving the lives of casualties. In either case, however, at times there is a degree of *apparent* random behavior brought on by the need to be ready to respond to any event in a given environment. My analysis follows along the lines suggested earlier by Kunimoto, *et al.*, to respect the phenomenological character of awareness. It also follows in the spirit of Polanyi [1967]: ". . by elucidating the way our bodily processes participate in our perceptions, we will throw light on the bodily roots of all thought, including man's highest creative powers."

First, some more preliminaries. Natural intelligence systems use the body to *attend* to physical things outside it. We attend to things outside our body, *from* our body, and we can feel our own body *in terms of* the things outside *to* which we are attending. There is a direction to the primitive relations of the preattention phase and of attending. There is also a sense in which we can make that external physical thing function as a *proximal term*[3] of the primitive relations of immediate awareness. That is, there is a sense in which *we extend our kinaesthetic bodily intelligence to include that object we attend to*, by extending our body to include the instruments, such as mechanical probes, we use to attend to an object or set of objects. Moreover, we do these things as well with purely abstract objects, including images. We can use images in our minds as proximal terms to probe or to stand in for physical objects.[4]

One who performs a complex task, such as playing a viola, tennis, or performing surgery, must know how to coordinate movements of their body, especially their arms, hands and fingers, by a kind of "cognitive indwelling" in the external physical thing that is functioning as a proximal term of their knowing how. In playing a viola, it is the viola itself, its strings and the bow used, and the music we are playing, the notes and our mental imagining of the music we play and hear, taken altogether, that form the proximal term. There is a sense in which all those become an extension of our body, specifically, an extension our bodily natural intelligence.

In the task of probing a wound, that proximal term is the interstices of that wound. One can only do this by using a refined discriminative sense of touch, gained over time and with much first-hand familiarity and experience, and mental imagining, to determine where the piece of shrapnel is located,[5] or *is likely to be* located, as well as to determine the extent of the injury caused by the shrapnel, and what must be done to remove it while saving the life of the victim. One's knowing how to do these tasks is *exhibited or disclosed* only in the *manner* of the actual doing of them. It is important to remember that by 'manner' I am referring to the dynamic refined sensitive touching and moving, the smoothly patterned oscillations, including timing, of one's body as the probing proceeds.

But there is far more. With respect to probing wounds, though there are similarities among such wounds, depending upon the overall wound presentation and ballistics of the weapons and ammunition which inflict them, each wound has it own unique signature, *unlike any other.* It has its own unique configuration of particulars. The term 'cognitive indwelling', Polanyi's phrase, is somewhat metaphorical language for the subject's primitive relations of immediate awareness of the object, the unique full configuration of the wound in relation to the body of the victim, functioning as a proximal term in the relation between the two. In this multilayered, highly complex relation, within a multifaceted complex of signs, the subject who knows how attends *from* his or her body, *from* the movements of their own hands and arms *and* tips of their fingers, *from* their mental imaginings that are continuously being formed as he or she proceeds, *to* the uniqueness of the configuration of those interior signs of the wound.

Part of such complex dynamic knowing how only *emerges* through time with much first-hand familiarity and practice and many errors, but aimed at a general concept of what a successful performance would be. Though we may be able to represent that general concept, the *distal term* of such *knowing how*, in a description, the *proximal* term(s) *from* which we attend to this general concept refers to those primitive epistemic elements of immediate awareness that can only be present with us. The proximal term of such knowing is inextricably linked to our bodily capacities, and is expressed or exhibited only in our actual doing. It is the proximal term of such knowing which discloses that we know more than we can say. In a sense, this knowing, found in the intersection of knowing how and knowing the unique, Boundary Set *S*, is an epistemic mean achieved between excess and defect in timing and seamless performance exhibited or disclosed in the somatosensory-motor manner of the actual doing of the task.

It is obvious that the exploratory movements of one playing a viola or probing a surgical incision or wound with one's fingers are guided in part by *continuously changing* present particulars or configurations of particulars from which one forms mental images. But these mental images are not solely those of what has been or is felt. They include images of what *has not yet been* touched, but is *anticipated.* If one is playing a piece of music with which one is familiar, one knows what to anticipate. But with respect to probing a wound, one must keep in mind that depending on the ballistics of the weapon and ammunition used, the trajectory of a piece of shrapnel or other projectile traced through flesh and bone is highly differentiated. The projectile itself changes as it traverses its path, flaying and twisting its sides outwardly, thus becoming more deadly but also creating a highly complex, unique configuration of its own in the wound it makes. One does not always know what to expect; but one must nevertheless know how to anticipate.

The one probing must *know how to anticipate the unique*, in part by attending *from* a multifaceted set of particulars felt by refined touch, *to* an image of how the remainder of a wound may be configured, based upon what one now feels from that refined touch. Moreover, the medical practitioner forms images of future events of the task, those images which include him or herself as participant, making choices as to which image he or she will bring into reality. These choices are based on those anticipations which in turn are based on *images of what has not yet been felt*, and also what has. Again, knowing how to probe such a wound cannot be gotten from *knowledge that* or from a set of explicitly formalized step by step procedural rules laid out from the start, though the performance as a whole must be highly informed by much *knowledge that* and many rules. Our knowing, as exhibited in such performances, is not exhausted by the distal term. In sum, this knowing found in Boundary Set *S* cannot be gotten by means of generalization, by knowledge *that*.

Before leaving surgical examples of knowing found in Boundary Set *S*, we should mention recent advances in laparoscopy. These can also serve to illustrate how technological advances in the development of physical probes have required increased medical training in immediate awareness and knowing how. This is especially the case to acquire first-hand familiarity, the "hands on" use of probes, in primitive relations of immediate awareness, especially *imagining, moving, touching*, to perform such tasks. Laparoscopy is an advanced technological technique that allows surgical instruments and a camera to travel into the body through small incisions. The advantages to the patient of such surgery are obvious: it is less invasive than traditional surgery thus reducing the level of all surgical and anaesthetic trauma to the body, faster recovery; such surgery allows a patient to leave a hospital within days. Traditional major surgery often requires weeks if not months of recovery. However, for surgeons to come to know how to perform laparoscopic techniques and to face virtually any eventuality in the operating room, surgeons must use direct, "hands on" practice with some unorthodox tools. There is no other way to acquire that knowing how than by entering those primitive relations of immediate awareness embedded within *knowing how*:

> Imagine trying to grasp an object with a pair of foot-long chopsticks. Think about doing this without looking at the object directly. Rather squint at the tip of each stick displayed in a picture on a color television. Finally, consider that the objects you are looking at are someone's gallbladder or spleen. Welcome to the hoary world of the laparoscopic surgeon.[6]

Training procedures for knowing how to perform this kind of surgery, as cited in the reference, include *timed* exercises which require a medical learner to use the nondominant hand to pick up black-eyed peas with a grasper. The miniature tweezers are attached to a long shaft, the end of which is hidden inside a box. The surgeon, who watches the position of these mechanical digits on a television screen must then manipulate the handle at the other end of the shaft to move the pea and drop it in a tiny hole. The point here is that one must not only do a great deal of *anticipatory spatially-related imaging* and timed, sensitive moving and touching following the patterns of action of the performance, one's sensory and somatosensory-motor system must become increasingly refined over time with such imaginary surgical experience. One's *knowing how*, based upon primitive first-hand familiarity with substitute surgical objects, must grow and become smoothly executed. In sum, with practice, one's *knowing how* emerges through time with the complex interactions of a very large number of epistemic (and other) elements or components.

In some ways, such medical performances, medical *knowing how*, bear a resemblance to a good tennis player. Borrowing a description from Ian Stewart:[7]

> Think of tennis players waiting to receive a serve. Do they stand still? . . Of course not. They dance erratically from one foot to the other. . .In order to be able to move quickly in any particular

direction, they make rapid movements in many different directions. A chaotic system can react to outside events much more quickly, and with much less effort, than a nonchaotic one.

There is a clear sense in which those who know how are poised on the edge of chaos or instability, ready to respond. They are poised on the boundary of order and disorder, and it is precisely this sense of 'boundary' to which I refer when I speak of kinds of knowing in Boundary Set *S.* I will return to a discussion of apparent random behavior shortly because apparent random behavior has had a profound effect upon our understanding of *rule-governed* as distinct from *rule-bound* behavior.

The point here is that though athletic games or performances such as tennis, and to some degree even certain surgical procedures, may be defined in terms of relatively simple rules independently of anyone actually doing them, natural intelligence systems such as humans who know how to play the game or how to perform the surgery, must behave in very complicated, dynamic, complex, self-organizing and adaptive ways which exhibit patterns of *knowing how.* Once one starts to analyze and examine that *knowing how* behavior in detail, *the simplicity of those rules is not there.*

Upon close analysis of the moves of a tennis player anticipating a serve, one finds primitive immediate awareness relations of the *preattentive* phase and *attending*, the primitive relations by which the player's sensory system continuously analyzes and filters stimuli, aligning sense qualities with stimuli. With these primitive immediate awareness relations, the player selects among possible objects in a field of perception to attend to the particular (or configuration of particulars) of the responding moves of one's opponent. Not only does one do the primitive selecting from all else, the background or "noise," the primitive relation of imagining is involved in one knowing how to spatially configure or arrange one's own body in anticipation of what one images as the move, or the likely move, the opponent will make in striking the tennis ball. Moreover, one must know how to recognize the configuration of moves of one's opponent and know how to anticipate which configuration will deliver the ball in a certain direction. One must also know how to ready one's immediate responses in the midst of a virtually immense, interrelated array of elements, while all along knowing how to handle the racquet (not too tightly with one's grasp and touch) and to respond to the spatial configuration and trajectory of a flying tennis ball toward one's end of the court, the background space in which one "frames" one's anticipated responses.

Again, there is a very large number of epistemic and other variables or components and their complex interactions involved in such a *knowing how* performance. There are not only the primitive epistemic relations and terms recognized by Russell in his *knowledge by acquaintance*, and the added primitive relations and terms of immediate awareness embedded within the patterns of *knowing how*, there are also the intricacies of the physiological and neurological systems.

Until very recently, scientific as well as philosophic inquiry were in pursuit of reductionist explanations of such complex phenomena. Classical science focused upon finding solutions to differential and partial differential equations which provide rates of change of elements over time. This was always a problematic approach with systems which were simply too large. However, the advent of the digital computer has assisted the geometrization of dynamic behavior of any complex system involving large numbers of complex interactions among large numbers of components. It provided the means of finding approximate solutions to dynamical equations very quickly. Instead of focusing upon precise numerical solutions to differential or partial differential equations, the focus has turned to what is called the *phase portrait* of a complex system. To understand how this approach, called *dynamical systems theory* can be useful in understanding Boundary Set *S*, I will initially introduce some technical concepts and methods.

5.2. A Framework for Thinking About Boundary Set *S*: Dynamical Systems Theory and Kauffman's[8] Random Boolean Nets for a Geometry of Knowing

Scientifically, there are fundamental differences between descriptions of organized simplicity found in closed equilibrium systems, such as a volume of gas or the actions of a pendulum, and descriptions of the organized complexity of open nonequilibrium systems, such as human knowing. Statistical mechanics provides us the means to obtain statistically averaged, typical and generic descriptions of simple, thermodynamically *closed* systems, such as a volume of gas at equilibrium. All the gas molecules obey the same Newtonian laws of motion and statistical mechanics. Thus such descriptions provide us an understanding of the averaged collective motions of those molecules. But natural organisms generally and human beings in particular are what are referred to as nonequilibrium thermodynamic *open* systems. Dynamical systems theory is used in those sciences which study living systems in general, from biological systems to human behavior, because they are highly complex systems in which there are very large numbers of interrelated parts ordered in highly complex ways.

The usefulness of dynamical systems theory as a framework for thinking and theorizing about Boundary Set *S*, as compared to thinking about simple closed systems, is that even though we may not know all the details of the order of interrelations among primitive relations and terms of immediate awareness and *knowing how*, we can nevertheless build a theory that seeks to explain the *generic* properties of the kinds of knowing found in the intersection of those sets.

Thus, I aim to characterize classes of properties of Boundary Set *S* that are typical or generic, and which do not depend upon knowing every detail. For example, in a cluster of networks of primitive relations of immediate awareness, we may not know where every proximal term is located (that is, every term of a possible immense number of terms of primitive relations), just as we do not know where every grain of sand is located in a desert. Indeed as applied to the neural basis of knowing, as Stewart has observed,[9] trying to locate a specific piece of neural circuitry in an animal's body is like searching for a particular grain of sand in a desert. Nonetheless, we can say a great deal about the properties of deserts and ice and neural circuitry. We can also say a great deal about the immediate awareness properties of a person knowing how to do simple to very complex tasks such as a *pirouette* or a surgical procedure.

As noted, I view the universe of knowing as *in part* a very large population of simple components, machines. In the development over time of a human knower, those components are not primarily *knowledge that* but are simpler components found in Boundary Set *S*. For immediate awareness, the epistemic primitives are the "species" of primitive relations identified earlier, minimally including primitives of the *preattentive* phase, *attending*, *sensing*, *imagining*, *memory*, *touching*, and *moving*. These are hierarchically ordered and related in very complex ways with differing kinds of space. The differing kinds of space are specifically related to the somatosensory *cum* motor relations of touching and moving to account for both distal and proximal terms of those primitive relations.

For *knowing how*, the epistemic primitives are largely the somatosensory-motor patterns of action which include timing and smoothly controlled oscillation of moving and touching. This population of simple machines, over time, constructs aggregates or clusters of simple rule-bound epistemic objects which interact and *transact* nonlinearly with one another and their environment to produce emergent structures. These emergent structures are forms and shapes of knowing, of natural intelligence. That is, they produce the behavior we all observe when we watch someone who knows how to do something as simple as tying one's shoes, dancing, performing a *pas de deux*, or as complex as playing a viola or conducting surgical probe procedures. The knowing found in Boundary Set *S* is a complex, dynamic, self-organizing, and emergent system.

To more precisely discuss kinds of knowing in *Boundary Set S* as the actual real time doing of tasks, I will first formally outline the larger formal theoretical framework within which Boundary Set *S* is found, demarcating the epistemic and epistemological universes, the universe of natural intelligence. This will provide an overview of what a *complete* theory would look like and how we obtain laws and law-like descriptions of it. In that overview, we will locate that part of it, the intersection of immediate awareness, *knowing the unique,* and *knowing how*, which I am addressing here. I will then introduce some technical concepts and methods useful in understanding the framework for thinking about the set of knowing found there.

5.3. The Formal and Geometric Structure of the Knowing Universe

To illustrate a dynamical systems approach to Boundary Set *S*, given that *knowing how* is exhibited in the manner of an actual doing of a task or performance, we must first distinguish the *epistemic* and *epistemological* universes. This will assist in setting forth an overview of a complete theory and the geometric structure of the part of it that I am addressing. Without repeating them here, I am assuming standard set operations, union, intersection, and so on, as well as the laws of set theory, De Morgan's, commutative, associative, and so on.

The epistemic universe, the universe of living intelligence, of knowing, is a set exhibiting very complex dynamic behavior. It is the set of intelligence. This is the object level of the theory. For our purposes, the *epistemic* universe consists in the following :

Knowing = Subject $\cup$ Object $\cup$ Content $\cup$ Context

On the other hand, the *epistemological* universe, or that universe of kinds of knowing, is a set consisting of the categories of knowing relations I earlier sorted as *knowledge that* [elsewhere termed *quantitative* knowing, abbreviated QN], performative *knowing how* [abbreviated PF], and *knowing the unique* [abbreviated QL]:

Ep = {*Knowing that* (QN) $\cup$ *Knowing the unique* (QL) $\cup$ *Knowing how* (PF)}

Epistemology, thus, is the study of knowing, focusing upon the nature of knowing itself and all kinds or categories of relations of knowing. This is equivalent to the study of intelligence.

I have also earlier identified the universe of signs as consisting of symbolic [SYM], iconic [IK], and enactive [EN] subsets.[10] We can distinguish the following universe of signs by which the universe of knowing, the epistemic universe, is representable, exhibited, or disclosed, that is *signed.*

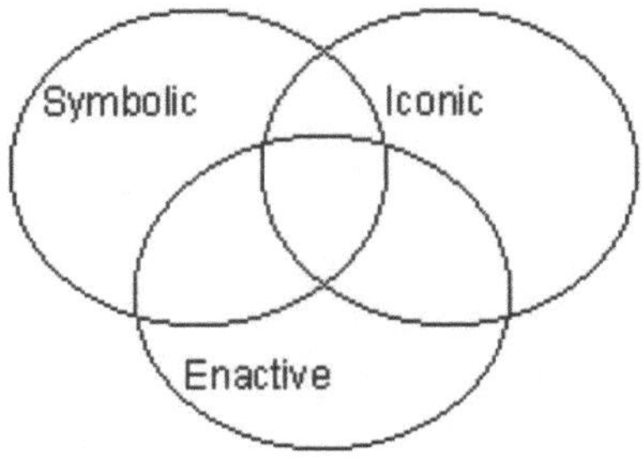

Figure FIVE-1. Categories of Signs

The epistemological universe is defined so as to focus upon both *signed* knowing, that is public expressions, representations, or artifacts of knowing, and *unsigned.* The *epistemic* set includes the above subsets, *subject, object, content, context*, and relations between them, and knowing may be either publicly signed or not publicly signed (for example, immediate awareness of an image or an object remembered).

A full and complete epistemological theory would minimally be concerned with properties of the two sets, the epistemological set and all its categories of relations of knowing, and the epistemic set consisting of the subsets *subject, object, content*, and *context* of knowing, and relations between them. It would also be concerned with the relations between the two sets, $\text{Knowing} \times \text{Ep} = \{(a,b): a \in \text{Knowing}, b \in \text{Ep}\}$ and the further relations between them and the set of signs, S, that is, $\text{Knowing} \times \text{Ep} \times \text{S}$. The number of subsets of $\text{Knowing} \times \text{Ep} \times \text{S}$ is the power set, $\wp$, of the crossproduct of those sets, $(\text{Knowing} \times \text{Ep} \times \text{S}) = 2^{|\text{Knowing} \times \text{Ep} \times \text{S}|} = 2^{|\text{Knowing}||\text{Ep}||\text{S}|}$.

It may be helpful to put these sets in some more formal perspective prior to a focus on the relations of knowing found in Boundary Set *S*. First, let me summarize some properties of the epistemological universe, the universe of kinds of knowing, in its relation to knowing.

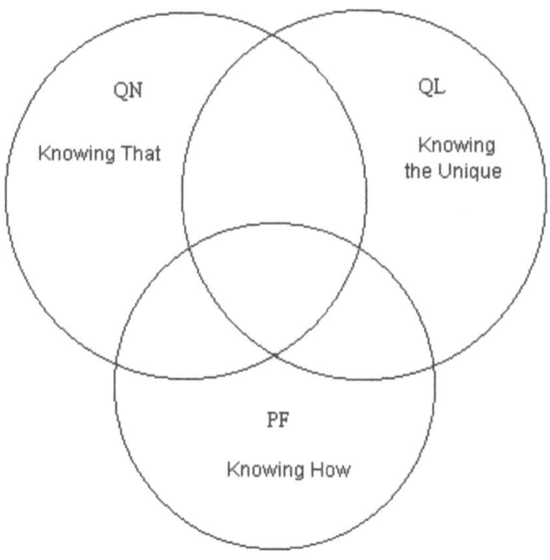

Figure FIVE-2. Epistemological Universe of Discourse

I view the *epistemic* or intelligence universe as a highly distributed, massively parallel knowing system, whether we are addressing natural or artificial systems which know and believe, whereas the *epistemological* universe may very well be serially computable, so long as it is addressed to public signs of knowing. The above classes of knowing are held to be related, relative to uncountable and countable or immense[11] domains, in the following way:

The public, *signed* [represented] expressions of each category are countable. The *unsigned* categories of *knowing the unique* are not countable, allowing for primitive terms of immediate awareness relations in transfinite domains.

The set of *knowledge that* represented in alphanumeric [symbolic] form is a countable domain. Coming to know *that* [not represented here] is a category not entirely countable because it overlaps with portions of the unsigned domains which are uncountable. *Knowing the unique* (QL) by itself is not countable because of the possible terms of immediate awareness relations in transfinite domains, though the primitive relations of QL are countable; those *signed* [exhibited, disclosed] portions which intersect with *both* knowing *that* (QN) and *knowing how* (PF) may be countable. *Knowing how* has portions which are not countable and portions which are. Those portions overlapping with QN are countable, and the portion overlapping with both knowing *that* and *knowing the unique* has portions which are countable and portions which are not. The part of the domain of *knowing how* with embedded *knowing the unique* within it, Boundary Set *S*, is not countable because of terms of immediate primitive relations in transfinite domains, and is a continuous domain.

Each of the above categories of knowing has been defined as relations of certain kinds between subjects and objects, with the focus here upon primitive epistemic relations of *knowing the unique*, QL, and the relations from QL to patterns of action and manner of *knowing how*, PF. A complete theory would set forth relations on the set *knowing the unique*, QL, that is subsets of QL × QL, and relations from QL to the set *knowing how*, PF, subsets of QL × PF. Because of the dynamic nature of these sets and their relations, directed graphs is a way to represent these relations. [I will discuss the use of random Boolean nets (or graphs) below].

As noted above, a complete theory would minimally be concerned with the two sets, the set of epistemological categories of relations of knowing Ep = {QN, QL, PF}, the epistemic set and its subsets, Knowing = {Subject, Object, Content, Context}, and the set of categories of signs of knowing, S = {SYM, IK, EN}. For example, from the Cartesian product of the sets Ep and S, we obtain nine ordered pairs:

$$\text{Ep} \times \text{S} = \{(\text{QN, SYM}), (\text{QN, IK}), (\text{QN, EN}), (\text{QL, SYM}), (\text{QL, IK}), (\text{QL, EN}), (\text{PF, SYM}), (\text{PF, IK}), (\text{PF,EN})\}$$

We could include in the universal epistemological set not only the categories of knowing but also the categories of signs of knowing. With this, we have the following epistemological set:

$$\text{Ep} = \{\text{QN, QL, PF, SYM, IK, EN}\}$$

We can then form the power set, $\wp$ of Ep, $\wp$(Ep), which is the set of all subsets of Ep, that is, given that for any finite set A with $|A| = n \geq 0$, $|\wp(A)| = 2^n$

$$\wp(\text{Ep}) = \{\ \varnothing, \text{E}, \{\text{QN}\}, \{\text{QL}\}, \{\text{PF}\}, \{\text{SYM}\}, \{\text{IK}\}, \{\text{EN}\}, \{\text{QN, QL}\}, \{\text{QN, PF}\}, \{\text{QN,SYM}\}, \{\text{QN, IK}\}, \{\text{QN,EN}\}, \ldots \{n_{64}\}\}$$

The formation of our power set $\wp$(Ep) yields 64 elements, that is possible classes of epistemological subsets. Knowing the power set provides us with information regarding the number of paths from one relation to another in a graph.

5.4. Digraph Theory of Knowing Relations

Not only would a full and complete theory be concerned with properties of the above three sets and relations between them, it would be concerned with setting forth at least the outlines of a hypothetical axiomatic material system. That axiomatic material system would include theoretical ordering by means of directed graphs or network theory, where epistemic and epistemological relations are directed graphs. Since my theory is directed largely to human and machine knowing phenomena, it cannot be a purely formal theory. A purely formal theory is one with only an abstract interpretation of the primitive terms. The theory concerns *contingent* relations in reality, thus requiring a physical interpretation. This presents some problems for checking the coherence of the theory, but given its partial formalization a check on logical consistency is nonetheless possible. This is so because there are deductive links to be checked out.

Ordering through digraphing gives us the advantage of presenting a theory which expresses relations which are contingent and also recursive and asymmetrical. That advantage is the use of path analytic techniques to check out correspondence of the relations expressed in the theory to those in reality. Path analysis is a procedure for estimating the path coefficients from correlational data using regression techniques.[12] More importantly, however, digraph theory provides a way to obtain *core, underlying properties* of epistemic and epistemological relations, specifically the relations of knowing found in the intersection of the sets in Boundary Set *S*.

The major advantage of directed graphs is that graph theory permits us to represent a relation as a graph. This same theory underlies neural network theory used for designing computer models for simulating kinds of knowing of interest to us. So, in principle, we can characterize very large numbers of primitive relations (and relations of relations of relations. . .) with the use of graphs of points and lines between them showing how they are connected (or not connected) with one another. Essentially, directed graph theory (or digraph theory) is mathematical theory which characterizes between pairs of points lines which can be directed. A digraph consists of a finite collection of points, P_1, P_2, P_3, . . .P_n, together with a described subset of the set of all ordered pairs of points which are directed lines. If D is a digraph, the graph obtained from D by removing the arrows is called the 'underlying graph' of D. I will set out just a few of the properties of digraphs.

A digraph D is said to be *connected* (or weakly connected) if it cannot be expressed as the union of two disjoint digraphs. This is equivalent to saying that the underlying graph of D is a connected graph. A *subdigraph* of a digraph is a subset of points and directed lines of the digraph which constitutes a digraph.

There are three different ways a digraph may be connected. A digraph is *strongly connected* or strong if every two points are mutually reachable. A digraph is *unilaterally connected* or unilateral if for any two points at least one is reachable from the other. A digraph is *weakly connected* or weak if every two points are joined by what is called a *semipath*, an alternating sequence of points and arcs (or directed lines) which are distinct.[13] Basically, a graph is an ordered pair, G = (V, E), with V = vertex set and E = edge set.

Epistemic sets can be represented on graphs as lettered points and a relation between two sets as a directed line segment (line segment with an arrow) connecting the two sets. Where there is an arrow or arrows between two points, there is a *directed connection* or *pairing.* A connection of one or more points or components to one or more other points or components is called an *affect relation.* Where there is a line without an arrow, a directed connection will be assumed in one or the other directions or in both directions. Where there is no line, there is no affect relation. An example is given below.

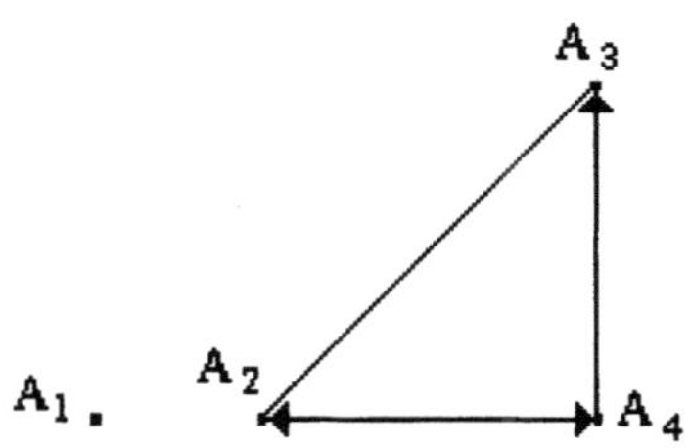

Figure FIVE -3. Example of Directed Graph

A *direct directed affect relation* is a directed affect relation in which the channel (line) is through no other point. An *indirect directed affect relation* is a directed affect relation in which the channel (line) is through other components (points). A completely connected epistemic affect system is, by definition, not possible since such a system would have complete connectedness if and only if all its epistemic affect relations were direct directed ones, that is direct channels from and to each epistemic component.

Where there is only one arrow between two points, the directed connection is *direct*. Where there is no line between a given point and other points, there is no connection or pairing with any of the other points.

A *directed walk* in a digraph is an alternating sequence of points and arcs (or lines). A *closed walk* has the same first and last points, and a *spanning walk* contains all the points. A *path* is a walk with all points (hence all arcs) distinct, except the first and the last. A digraph is *unilateral* if and only if for every pair of distinct points, P_x and P_v, there exists a directed path from P_x to P_v or (in the exclusive sense of 'or') from P_v to P_x. It follows that all strong digraphs are unilateral, however it is not the case that all unilateral digraphs are strong.

A digraph is *disconnected* if and only if the points constitute two disjoint digraphs, that is with no directed line joining any point in one digraph to any point in the other. Digraphs, hence, fall into four mutually exclusive subsets with respect to connectedness:

C_o = disconnected digraphs
C_1 = weakly connected digraphs
C_2 = unilateral digraphs
C_3 = strongly connected digraphs

The following theorems obtain showing the bounds on the number of directed lines in a digraph in each of the above categories and have been used to show the significance of graph theory to epistemological theorizing in general:[14] Where

'D' = digraph
C_1, C_2, C_3 = subsets of digraphs with respect to connectedness
p = number of points
q = number of lines

then the following theorems (T) obtain:

T_0 = If D is in C_0, then $0 \leq q \leq (p-1)(p-2)$
T_1 = If D is in C_1, then $p-1 \leq q \leq (p-1)(p-2)$
T_2 = If D is in C_2, then $p-1 \leq q \leq (p-1)^2$
T_3 = If D is in C_3, then $p \leq q \leq p(p-1)$

5.5. Properties of Relations: Natural and Artificial Intelligence Systems

By using the properties of directed graphs, one can determine the properties of relations. A relation is *reflexive* if every V has a loop, that is an edge to itself. A relation is *irreflexive* if no V has a loop, that is there are no loops in the graph. A relation is *symmetric* if All E are in pairs and/or loops, that is if the graph is undirected. A relation is *antisymmetric* if each E (except loops) is directed only one way. A relation is *transitive* if there is a path $x \rightarrow \ldots \rightarrow y$, then there is an edge $x \rightarrow y$. A strongly connected graph represents a transitive relation because every vertex has an edge to every other vertex. An *equivalence* relation on a graph and digraph is reflexive, transitive, and symmetric. That is, the digraph is all loops, undirected, and within a connected graph. Equivalence relations look like a collection of complete graphs plus loops.

A *partial order* on a graph is one which is reflexive, that is it has loops, it is antisymmetric, that is it is directed, and it is transitive. With partial order, any two things x and y that the graph is intended to represent are either related or they are not. A set A is *totally ordered* if for all $x, y \in A$, either xRy or yRx. That is, everything is comparable. Where sets are defined over Z, the integers, everything is comparable if in the right order [$\leq$ = total order, and only applies to Z if the elements are in the right order; applies to the reals, R, but not to the complex numbers, C]. On the other hand, the subset relation, $\subseteq$, is partial order because one cannot compare elements. Where one has a partial order on a finite set, one can extend this to total order by doing a topological sorting[15] because total order is a subset of partial order.

Through axiomatization and digraphing, because they are both ways of ordering explanatory theoretical sentences, we have evidence of completeness. Any gaps in the theory will be shown because missing deductive links will be apparent in axiomatization and missing connections will be apparent in the case of digraphs. With respect to digraphs, presented as path diagrams meeting the requirements for path analysis [connections must be asymmetrical], the density and connectedness of the digraph indicate whether connections are missing. Density is the number of direct connections over the number of possible connections, given by the following equation:

$$D = DC/N(N\text{-}1)$$

Where 'D' stands for density
'DC' stands for number of direct connections
'*N*' stands for the number of properties

Density cannot fall below some minimal value because obviously less than *N*-1 direct connections results in some properties not being connected. (On the other hand, as I will discuss a bit later, if the density of a network of connections is too great, it will result in chaos).

Connectedness is the number of direct and indirect connections over the number of possible connections, given by the following equation:

$$C = DC + IC/N(N\text{-}1)$$

Where 'IC' stands for the number of indirect connections.

I will return to the use of density and connectedness below in a discussion of information theory and random Boolean graphs of relations between primitive epistemic relations of *knowing the unique* and patterns of action of *knowing how*.

The Intersection of *Knowing the Unique* (QL) and *Knowing How* (PF), QL ∩ PF

In the theory of Boundary Set *S*, the focus is upon the sets (QL) and (PF) between which certain elements are related in some way. For example, we may want to focus upon the set of ordered pairs of epistemic conditions of a certain category satisfying a certain equation.

This set would be a subset of QL × PF and would contain all points of the graph of the equation. The subset is a relation from QL to PF. In general, for finite sets A, B with |A| = *m*, and |B| = *n*, there are 2^{mn} relations from A to B, including the empty relation as well as the relation A × B itself. There are also 2^{nm} relations (= 2^{mn}) from B to A, one of which is also ∅ and another of which is B × A. Formally, a subset Y ⊆ A × B is called a relation from A to B (a binary relation, since two elements $a \in$ A and $b \in$ B are involved). [In some mathematics texts, the symbol '~ ' stands for the phrase 'is related to'. Thus, the statement $(a,b) \in$ A is written $a \sim b$, which means *a* is related to *b*.]

A set A and a partial order ≤ together form a pair (A, ≤) called a partially ordered set. Where $a \leq b$, *a* is said to be smaller than or precede *b* and *b* is greater than or follows *a*. Where $a \neq b$, then $a \leq b$ is written as $a< b$ (with $a, b \in R$). A partially ordered set (A, ≤) is totally ordered if, for every $a, b \in$ A, either $a \leq b$ or $b \leq a$.

There are other generalizations of graphs which should be mentioned here as they will be referenced briefly later in the use of random Boolean graphs or networks to characterize the dynamics of primitive epistemic relations. I have shown that graphs are used as models of relations. We could also assign weights to edges in a graph to represent degrees of relations, selecting possible values in advance. There are also graphs of systems of differential equations which can model real physical flows of things like electricity or water. And the use of graphs of systems of difference equations can show us stability properties of things. For example, by graphing the equations, we can ask, "Do solutions to the equations stabilize?" Do they grow without bound? Do the equations show an oscillation?

Ultimately, what we are striving for is a focus more upon the *geometry*, the forms and shapes of the dynamics of a system of elements, rather than upon *symbols* of equations. With complex dynamic systems, such as human knowing, especially in Boundary Set *S*, we cannot even keep track of the equations since the computations are sometimes of a very large number of elements. We simply have to watch as the dynamics of the equations unfold on a computer screen.

We are not necessarily assuming *natural* knowing systems here, since *artificial* systems exhibit certain of the kinds of dynamics and behavior of interest. We assume that the epistemic universe as a whole has a very complex geometric structure, and that the set of points making up the boundary of that set (and indeed the boundaries of the subsets within the set) exhibits a very rich and extraordinarily dynamic and complex structure, (which I will later show to be on analogy with the Mandelbrot and Julia sets).[16] The latter will be used to illustrate certain properties of the epistemic universe as a whole and Boundary Set *S* in particular to show the serious limitations on the classical computational/decidability approach to such sets defined over the integers.

After the introduction of a few more technical concepts necessary to understand a dynamical systems approach to knowing systems, especially Boundary Set *S*, I will return to a more detailed explanation of the use of directed graphs, that is random Boolean networks, to represent epistemic relations.

Vectors, States, and Trajectories

A vector is an ordered set or list of variables. In the design of an intelligent machine, one must describe many variables and characterize many simultaneous multivariate computations. Thus, vector notation is one way to do this.

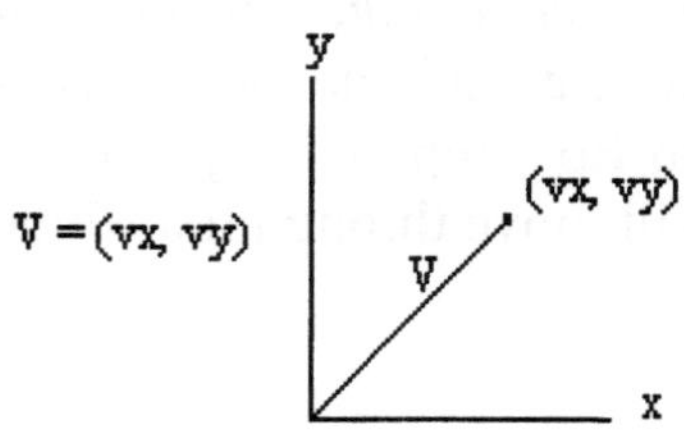

Figure FIVE-4. Vector

Components of a vector can be coordinates of a point [for example above, (vx, vy)] corresponding to the tip of the vector. A vector space is defined as the locus of all pairs of components that can exist. Vectors can have two or more components. A vector with two components is a *surface*; three components defines a *volume*; four or more components defines a *hyperspace*. A hyperspace can include an immense (hyperimmense) or *googol* and *googolplex* number of elements, which are finite and countable, *but are so large they cannot be handled by conceivable computational methods*. Recall the Berry Paradox. This involves the paradox of human beings nonetheless *naming* such large numbers with which they cannot possibly be acquainted, immediately aware, even if they had the lifetime of the universe to count them.

Vectors can specify a state which is an ordered set of variables. For example, the epistemic state of a person or machine might be characterized by the following state vector: Ep = (ep_1, ep_2, ep_3), where:

Ep_1 : *knowledge that* [quantitative knowing, QN]
Ep_2 : *knowing the unique* [immediate awareness, QL]
Ep_3 : *knowing how* [performative knowing, PF]

A given vector might also be a list of epistemic primitive relations translated into real values. The above state vector is in a space consisting of all possible combinations of values of the variables in the ordered set (ep_1, ep_2, ep_3), defining the space S_{ep}. Every point in that space corresponds to a unique epistemic (knowing) condition, and the entire space corresponds to all possible epistemic (knowing) conditions. In terms of obtaining of a given knower, each variable of *knowing the unique* (for example relations and terms of *attending*, *imagining*, *moving*, and *touching*), *knowledge that*, and performative *knowing how*, is time-dependent. Thus we can add one more variable, time (t) to our state vector summarized as Ep = (ep_1, ep_2, ep_3, t). That is, through time, the point defined by Ep will move through four-dimensional space.

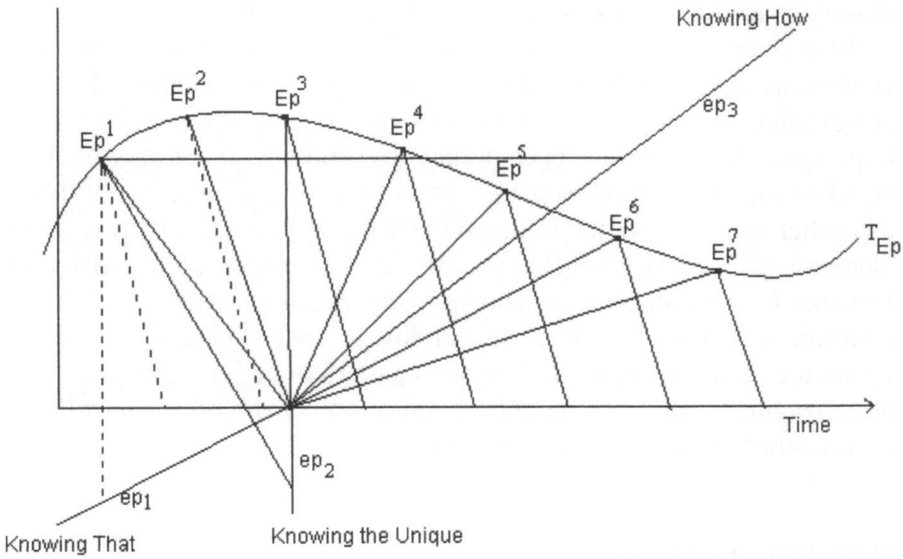

Figure FIVE-5. Trajectory of Knowing

The above figure shows the locus of the point traced by Ep as it defines a trajectory, T_{Ep}. For our purposes, I have added the variable, time (t) to our state vector. Through time, the vector will move along the time axis. Since we take each of the other variables as time-dependent, the trajectory will not be a straight line parallel to the time axis, thus the trajectory will be some curve.

It is important for our purposes in the analysis of Boundary Set *S*, that many things can be represented as vectors, including *signs* which function indexically. For example, gestures and motions or patterns of action which "point" can be represented by a trajectory, such as the manner of a person's actual doing of the task, probing a wound. Also, pictures or imaginings can be represented as two-dimensional arrays of points, each with its own brightness and hue.[17] Each point would then be represented as three numbers corresponding to color brightness (red, blue, green). Where each of the numbers is zero, the color is black; where they are large and equal, the color is white. Where there is a large number of points spaced closely together such that the human eye cannot discern the spaces, the eye cannot distinguish that array of a closely spaced large number of points from a real object.

Moreover, sounds, musical notes and chords can be represented as vectors, as well as symbols or signs. Any ordered set of binary digits corresponds to components of a binary vector. Thus if the underlying hyperspace is continuous, each point corresponding to some symbol or sign has a neighborhood of points around it which are closer to it than any other symbol's points. In much the same way, we can represent the elements of Boundary Set *S* as vectors. That is, the underlying space of *knowing how* in intersection with *knowing the unique* may be continuous with each point corresponding to a set of primitive epistemic relations embedded within a pattern of action. That point is an *attractor*, which has a neighborhood of points around it closer than to any other element of the hyperspace.

Functions and Operators

Within the context of vectors, a *function* is a mapping of points in one hyperspace onto points in another. In mathematics generally, a function is a relationship between symbols which can sometimes be in a one-to-one correspondence with physical variables. Where the relationship is in a one-to-one correspondence with physical variables, there is sometimes an asymmetry, "sense" or direction in the relation, to use Russell's phrase. For example, we can map a set of states [that is, a state vector, or all combinations of values of variables in an ordered set] defined by independent variables, that is a set of *causes*, onto a set of states defined by dependent variables, a set of *effects*. This can be expressed as: f: C→ E. This means that f is a relation mapping the set C into the set E, that for any particular state in C, f will compute a state in set E. Functions can be expressed in a variety of ways: as an equation, as a graph, as tables or matrices, and as circuits.

(i) $x = f(y)$ is an equation which reads 'x is a function f of y'.

(ii) $x = 2y^2 + 3y + 6$ is an equation expressing a relationship between x and y.

(iii) The following graph represents the function also expressed as an equation:

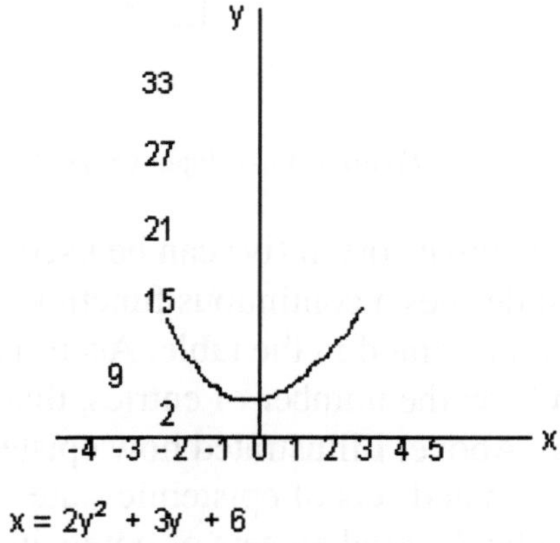

Figure FIVE-6. Graph of a Function

A function can also be expressed as a matrix or table as follows:

(iv)

z	*x*	*y*
0	0	0
0	0	0
0	1	0
1	1	1

Information in the above matrix or table can also be presented in the following figure:

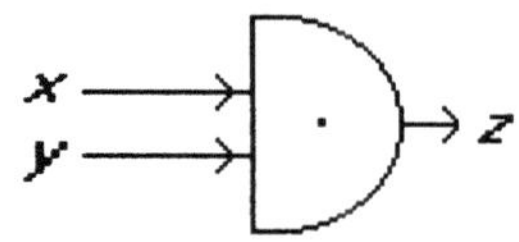

Figure FIVE-7. Input-Output Graph

Tables of the above sort in (iv) can be used to define non-Boolean functions, but it defines a continuous function only at the discrete points that are represented in the table. Accuracy of a continuous function depends on the number of entries, that is the resolution on input variables. Above, I illustrated how epistemic states can be denoted by vectors and sets of epistemic states [epistemic relations and their terms] can be denoted by sets of points in hyperspace. I will extend the notion of a function as a mapping from one set of states to another to a mapping of points in one vector hyperspace onto points in another.

An *operator* can be defined as a function mapping input Ep = (ep_1, ep_2, ep_3, . . .ep_n) onto the output scalar variable K [*knowing*], written either as K = H (Ep) or as K = H (ep_1, ep_2, ep_3, . . .ep_n). The functional operator is often indicated by engineers as a circuit or "black box", which exhibits input-output processes.

Where we have a set of operators, h_1, h_2, h_3, . . . h_n operating on an input vector, Ep, as in Figure 7, we then have a mapping H : Ep $\rightarrow$ K [alternatively, K = H (Ep)] where the operator H = (h_1, h_2, h_3, . . . h_n) maps every input vector Ep into an output vector K. Ep is a vector or point in input space. The [information] function H is a mapping from input space onto output space.

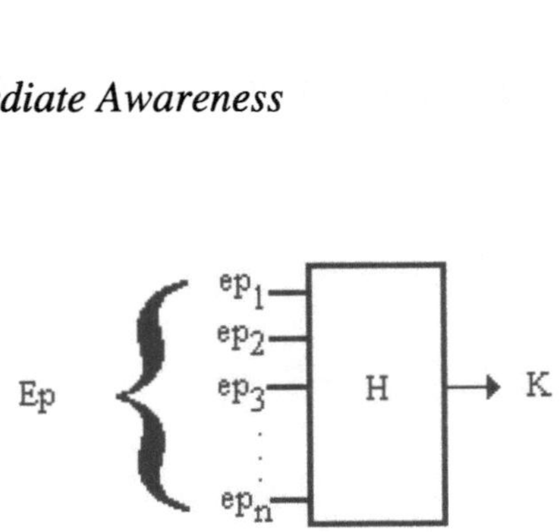

Figure FIVE-8. H Function Map of Input Ep into K

In the above Figure, H is a function mapping input vector Ep into scalar variable K. (H denotes information functions).

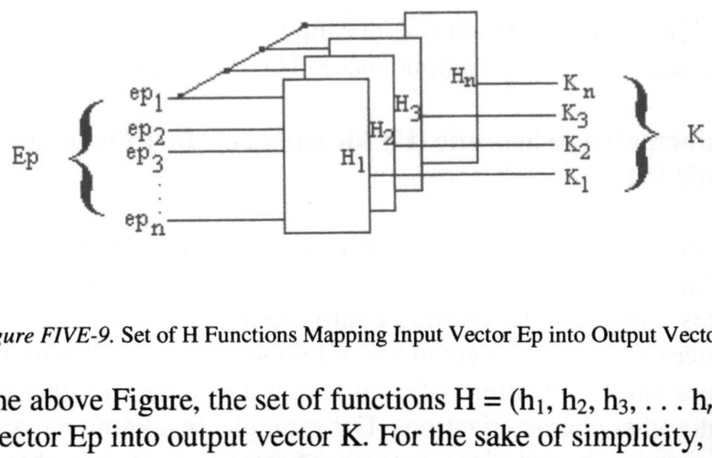

Figure FIVE-9. Set of H Functions Mapping Input Vector Ep into Output Vector K

In the above Figure, the set of functions H = (h_1, h_2, h_3, . . . h_n) maps input vector Ep into output vector K. For the sake of simplicity, I will limit our discussion to a single Ep which will map into one and only one K. As we stated above, variables in Ep may be time-dependent, hence Ep will trace a trajectory T_{Ep} through an input space. The H function will map each point Ep on T_{Ep} into a point K on a trajectory T_K in output space.

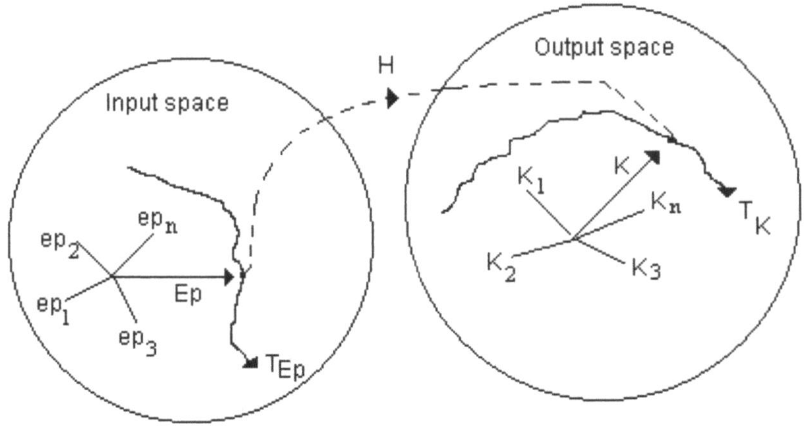

Figure FIVE-10. H Mapping T(Ep) into T(K)

In Figure *FIVE-10*, H maps every input vector Ep in input space into an output vector K in output space. Thus, H maps T_{Ep} into T_K.

5.6. Information-Theoretic (H) Measures of the Universal Epistemic Set

Information theory is a probabilistic theory or model of information or uncertainty. It is theory which characterizes occurrences or events at categories. For example, epistemic information is a characterization of occurrences at epistemic categories of knowing systems. Moreover, sometimes characterizations of occurrences are sometimes themselves made into other characterizations. For example, in a knowing system, a decision can be characterized in terms of categories of natural language expressions, which in turn can be characterized in terms of dots, dashes, and spaces of Morse code.

The broader term 'information' takes on two different senses depending upon whether there are *alternatives*. There is the *non-selective sense* of information in which there are no alternatives. For example, in the characterization 'H_2O is the formula for water', there are no alternatives. The characterization [or "situation"] is fully specific and there is no uncertainty. From a non-selective point of view, such as found in Hartley's[18] early work, there was information, even though there was no uncertainty. However, from a *selective sense* of information, there is no information. Thus in the early sense of the concept, the concepts 'information' and 'uncertainty' were not equivalent. But in the selective sense, there must be uncertainty in the characterization or situation for there to be information. In the characterization, 'Either smoking is related to cancer or it is not', there is an alternative. It either is related or it is not. Thus there is information. There is uncertainty because not all occurrences can be characterized by means of only one category. This is the sense of information, defined in terms of uncertainty, that is of interest to us in any information-theoretic characterization of Boundary Set *S*.

In general, our uncertainty in some situation is reduced by an action [an occurrence or event]. Thus the occurrence may be viewed as a source of information, and the amount of information obtained by the action can be measured by the *reduction of uncertainty* resulting from the action. This sense of information is defined strictly in a formal, syntactical sense. It is not the sense of information usually found in ordinary human natural language contexts. Nonetheless, we can view communication in a more general sense as a characterization of information-theoretic measures of a natural intelligence or knowing system in general, as well as between humans, between knowing systems generally.

Viewing natural intelligence as a system, information theory can give meaning to the categorization of the components and connections and interconnections (interrelations) of that system. Every system has information in the sense that occurrences of its components, relations, or relations of relations can be classified according to categories. The added condition of selectivity of the information, uncertainty of occurrences at the categories, is required to develop information properties of systems (and negasystems, what is not system) and of their states.

To illustrate the use of information-theoretic measures in the characterization of a knowing system in general, and Boundary Set *S* above in particular, we use the set of epistemic categories above to show how the measures would work. We will address the problematic nature of immediate awareness, consisting of the "species" of primitive relations noted earlier. For now we will set forth each category in set-theoretic terms.

If we took into consideration all the elements in the set of QL as well as performative knowing PF, we would have the following: QL = {Pre-A, S, M, I, F-ha, R}. These include the primitive relations between a subject, S, and object, O, in addition to touching and moving, shown in Figure *FOUR-5* in the last chapter. Intentional touching and moving are included in First-Hand Familiarity. Moving and touching are viewed as in some respects as *meta*-level primitive relations, consisting of primitive sensory-motor relations in combination with the other relations of imagination and memory. These are all, of course, "layered" upon preattentive and attentive primitive relations which themselves consist of large numbers of primitives. Obviously, in the dynamics of immediate awareness there are relations among these relations and relations of relations of relations, *ad infinitum*. Measures of these can quickly cause combinatorial problems, but at this point we just wish to get intuitively clear on the overall formal theoretical framework within which we must approach Boundary Set *S*.

The power set of the above set , QL = {Pre/A, S, M, I, F-ha, R}, would be gotten by $2^6 = 64$ elements. If we sort the set of knowing how, PF, as having 4 elements consisting of kinds of performance sorted according to number of paths and termini, then we have the set of performative knowing PF as follows:

PF = {Pr, Co, In, Cr}[19]; the power set of this set = $2^4 = 16$ elements

The Cartesian product of *n* sets $y_1 \times \ldots \times y_n$ = set of *n*-tuples whose $_i$th members all belong to y_i : $\forall y_1 \ldots y_n (y_1 \times \ldots \times y_n = \{ < z_1, \ldots z_n >: z_1 \in y_1 \,\&\, \ldots \,\&\, z_n \in y_n\})$

From the Cartesian product of the above sets, assuming QN has 8 elements, we have:

QL = {64} × QN = {8} × PF = {16} = 8,192 ordered pairs.

In an extended study, we would want to explore various measures of *uncertainty* as they relate to all of the epistemic and epistemological categories. In most studies that I am aware of, the use of information theory applied to theories of knowledge has largely been restricted to measures of belief and plausibility dependent upon evidence, as we see with Shafer [1976]. Those measures provide us with the means to measure uncertainty with respect to knowledge *that* or belief *that* claims, which are not of concern to us here. Information measures of knowledge *that* are relatively non-problematic as long as we have linguistic or alphanumeric propositions representing a system's knowledge *that* and evidence as well as possibility and probability measures.

5.7. Mechanism or Organicism

But an information-theoretic approach to kinds of knowing in Boundary Set *S* presents unique problems. In part, the problems stem from how we conceptualize the human knowing system, or natural intelligence system, to begin with. Models of systems can be classified as to whether they are mechanistic or organismic. A mechanistic model or point of view is one in which the object of inquiry is represented like a machine. A machine is an object consisting of parts that act in predetermined ways. (I am also using the term 'machine' as equivalent with 'algorithm'). In machines, the parts have natures that are not alterable and have fixed actions. And the actions which are specific to a certain kind of machine result from a combination of the parts. Generally, in a mechanistic model, the emphasis is going to be on the parts which are taken as non-modifiable and as determining factors. Simple input-output-feedback models and classical stimulus-response models, are mechanistic. They are the closed systems of classical physics. That means they are isolated from their environment.

On the other hand, an organismic model is one in which the object of inquiry is represented like an organism, a living thing. An organism is one that is a structured whole, in which the content and form of its parts are determined by its function. Thus, the parts are alterable; they do not have fixed actions. Rather, the parts act interdependently to maintain function and wholeness. The content and form of the parts change relative to a whole. In an organism, the emphasis is on the whole as determining its parts. Organisms are open systems because of the information-theoretic extensions into their environment. They interact, transact, and act on and with their environment. Those uniquely human kinds of knowing found in Boundary Set *S* must be viewed on the model of an organism, not a machine.

To try to visualize how digraph theory and information theory come together, try imagining at first a vast, multilayered, densely interconnected set of "sheets" or tapestry made up of billions of threads of varying weights, strengths, and sizes. Visualize them as inter-woven together; some of the threads are strongly woven together and knotted, while others are not as strong. The threads are the lines connecting the points in digraph theory. The knots may be neurons and threads the connections between them. Digraphs give us a way of visualizing and theorizing about all those *billions* of neuronal connections in a single brain. It is those billions of connections which permit us to be immediately aware of immense numbers of primitives in the preattentive phase, that in turn get deployed in attention, and in touching and moving of our knowing how to do something. Attention is another level of awareness, when our "awareness *that*" may or may not kick in, as we are continuously and dynamically interacting, transacting and acting with and upon our environment, our ideas, images, tastes, smells, desires, and whatever else may be of interest to us. In principle, information theory lets us measure every single one of those connections when something occurs and the connection gets reinforced, the knot tightened.

We have to start however, with one important assumption. We assume that human knowing in Boundary Set *S* (and natural intelligence systems generally) is not complete or "all knowing." Hence there *must* be *uncertainty* of occurrences at the categories where we interpret that uncertainty in terms of probability distribution. The existence of alternatives for the occurrence of any epistemic component, such as a primitive relation of immediate awareness, indicates the selective sense of information. This can be measured and such measures of the *transmission* of information shared between a system and its environment or between two or more systems can be calculated with the Shannon [1949] formula.

The basic information function is designated by 'H', as in the above graphs. By summing over the amount of information associated with each selection, weighted by the probability that the selection will occur, the value of H can be obtained. To state this more precisely, H(C) is the average uncertainty per occurrence with reference to the classification C. *It is the average number of decisions needed to associate any one occurrence with some category c_i in C,* with the provision that the decisions are appropriate; it is a function of the probability measures in C:

$$H(C) = -\sum_{i=1}^{n} p(c_i) \log_2 1/p(ci)$$

A measure of joint uncertainty would be:

$$H(C_{IJ}) = \sum_{i=1}^{m} \sum_{j=1}^{n} p(c_i, c'_j) \log_2 1/p(ci\ c'_j)$$

The measure for conditional uncertainty would be:

$$H(C_I\ C_J) = \sum_{i=1}^{m} \sum_{j=1}^{n} p(c_i c'_j) \log_2 1/p(ci\ c'_j)$$

The three H measures are related as follows:

$$H(C_I) + H(C_J\ C_I) = H(C_{IJ})$$

The T measure is the amount of shared information:

$$TIC_I,C_JP = H(C_i) + H(C_J) - H(C_{IJ})$$

If we apply all this to neural networks in computers, we can also use these information measures with measures of Hebbian learning so as to obtain a portrait of *incremental* coming to *know how* over time. The usefulness of these measures in the context of random Boolean graphs of Boundary Set *S* will become more apparent in later discussion. Just keep in mind that information in every occurrence at a category of primitive relations can in principle be measured with one or more of the above measures.

But there is one important difference to note between information measures available on machine models of things and those measures on organismic models. Machine models are limited in their information measures because of their isolation from whatever constitutes their environment. On the usual machine models, input is what the system takes in. But measures of what is available to the system from the environment and how that relates to the input remains unknown. Also, output on the usual machine model is what is available from the system, but what the environment gets and how that relates to the output also remains unknown. Feedback on the usual machine model simply relates the output to the input. Basically, with machine models, information directed toward the maintenance of specified goals is analyzed, while information not in the system is excluded from analysis. By their very nature, machine models cannot provide information necessary for self-organizing complexity and adaptive properties found in living things.

5.8. Poincaré Map and Random Graphs of Primitive Knowing Relations: From a Symbol-Based View to a Geometric View

Above I made reference to a space of all possible combinations of values of variables in a system. This concept of a space of all possibilities is an invention of Henri Poincaré, called a *Poincaré map*, and is used in dynamical systems theory to refer to the range or space of possible behaviors. Poincaré's conceived the idea to draw a picture that shows what happens for all possible initial values of a system. The vertical axis might correspond to the knowing condition of a person, with the horizontal axis corresponding to the interrelations of kinds of epistemic variables (primitive relations) possible to obtain of a knower. A point on the map will correspond to possible combinations of conditions (*e.g.* the primitive relations of immediate awareness and patterns of action of *knowing how*) possible to obtain of the knower. The point through time will trace a curved trajectory of knowing.

We can start the map with different initial values, getting any number of different curves. A complete set of such curves flowing on the plane is called the *phase space* of the system and the curves themselves are called the *phase portrait*. Instead of a *symbol-based* idea of a set of differential equations with various initial conditions, we have a *geometric, visual* scheme of points flowing through epistemic, natural intelligence, space.

If we look upon a person who knows how to perform any of the tasks or performances found in Boundary Set *S*, such a map will represent all possible patterns of action and motions (for example, refined moving and touching) of that *knowing how*. It will also represent all possible primitive epistemic relations of *knowing the unique* exhibited by the manner of *knowing how*, and all possible combinations of relations between these two sets.

That is, a given point will represent all possible combinations of such epistemic values for a knower who knows how. Obviously, the Poincaré map is a theoretical hyperspace map of conceivably an immense number of dimensions since it represents a space of all mathematical possibilities of combinations of elements found in the intersection of the two sets. The map provides us with a concept of the *epistemic space* or specifically the Boundary Set *S knowing space* over which such a knowing system may traverse. If we think of knowing systems, especially those found in Boundary Set *S*, as networks or directed graphs of relations, then that system can assume a vast number of possible knowing states. *We can think of a knower as traversing across this mathematical landscape of possibilities*.

As represented above, a moving point or trajectory of *knowing how* traces a curve which is a visual representation of the future knowing behavior of the knower. By examining the curve, we can determine very important core, fundamental features of the dynamics of the knowing itself without being concerned with actual numerical values of the coordinates. If the curve closes into a loop, for example, then we know that the epistemic variables are following a *period cycle*, they are repeating the same values over and over again. If the curve seems to hover toward some particular point and stops, then the knowing system has settled down to a *steady state*.

As Stewart explains,[20] the significance of the Poincaré map is that dynamics can be visualized in terms of geometric shapes called *attractors*. Starting the system from some initial point and watching what it does in the long run, any system generally ends up wandering around on some well-defined shape in phase space. That shape is an attractor and the long-term dynamics of a system is governed by its attractors. This fits intuitively very well with the way we think of intelligence generally and knowing in particular. Where attractors are interpreted as kinds of knowing, these generally govern the long-term dynamics of a person as a natural intelligence system, as a knower.

If a system settles down to a steady state it has an attractor that is just a point. It is highly unlikely that a person as knower has only one attractor. Most likely, such a system has quite a few since most people know how to do quite a lot of things and they know how to do them simultaneously. A system settling down to repeating the same behavior periodically has what is called a *closed loop attractor*. Closed loop attractors correspond to oscillators and it is those in particular that are of interest to a dynamical systems analysis of *knowing how* in intersection with *knowing the unique*, where we are focusing upon bodily kinaesthetic tasks. Recall that *knowing how* is defined in terms of *manner of performance*, with *manner* defined in terms of timing and smooth patterns of moving and touching. Paraphrasing Stewart,[21] *knowing how* hooks together huge circuits of attractors, that is oscillators, which interact with each other to create complex patterns of *knowing how* behavior.

Complex systems in general have attractors which are just whole families of trajectories, to which the systems settle down. These attractors or families of trajectories can be interpreted in a variety of ways. For example, in immune systems, the attractors can be interpreted as different immune states. McClelland and Rumelhart[22] interpret alternative attractors in neural networks as alternative memories or categories by which the network "knows" its world. Their interpretation fits along the same lines of my interpretation of Boundary Set *S*. Alternative attractors can be interpreted as epistemological categories I sorted above, *knowledge that*, *knowing the unique*, and *knowing how*. For purposes of illustrating properties of Boundary Set *S*, however, I limit the interpretation to the latter two categories.

The image of knowing in Boundary Set *S* which I have tried to present so far is of a highly complex system of vast networks of interrelated primitive relations. A key to understanding the dynamical systems approach to natural systems exhibiting *knowing how* and *knowing the unique*, as I indicated above, is the following, paraphrasing Kauffman[23]: If all properties of natural knowing systems depend on knowing every detail of their relational structure and logic, if natural knowers are arbitrary widgets inside arbitrary contraptions all the way down, then the epistemological problems are not just vast. They would in effect be impossible given the virtually immense (if not uncountably infinite) number of possible relations and terms of those relations. We would have to know all the details to understand any of it.

5.9. A Toy Model of a Random Graph: Kauffman's Buttons and Threads for a Tapestry of Knowing

To understand more fully the core properties of natural knowing systems, we might find it useful to expand on my "thread" or tapestry model above. We can imagine a person who knows how to do a lot of the tasks found in Boundary Set *S* as a network (or a system of interrelated networks) of primitive relations. Following Kauffman's[24] toy model, we could imagine each primitive relation of immediate awareness and the relations of *knowing how* as buttons, with threads or strings tied to each button representing the connections or paths between them. If we imagine the knower in a state prior to knowing anything at all (that is, without an epistemic relation between the subject and an object), we might assume the buttons are randomly scattered about on some surface.

It is theoretically useful to view systems from both a random as well as a nonrandom perspective (where the parameters are some constant). If we have a large number of buttons (relations) scattered about, we could then randomly choose two and connect them with a thread. Putting that pair down, we could then randomly choose two more, connecting them with a thread and also putting them down. If we continue to do this, we will inevitably randomly pick up a button that we have already picked up before and which is already connected by thread to yet another button. Thus, when we tie a thread between the two newly chosen buttons, we will find three buttons tied together, a triple relation. Continuing on with our random choices and tying the buttons (relations) with threads, we inevitably end up with an interconnected large cluster, *assembly*, or circuit, ensemble of buttons and threads. Among other things, we have increased the diversity of the relations. From that cluster, if we at random pick up a given button (relation), it will be interconnected with many others, though we may still have buttons not connected with any others.

The buttons and threads model fits fairly well with Hebb's idea of a cell assembly. He proposed that individual neurons or groups of neurons would spontaneously hook themselves up to form reverberating networks. Neuron A would activate neuron B, which activates neuron C which then reactivates neuron A again. This toy model also fits very well with the view Russell had of knowledge by acquaintance. He once commented[25] that many problems in philosophy require the consideration of *triple*, *quadruple*. . relations, which have in general been unduly neglected.

The toy model is an example of a *random graph* or network, where the buttons as relations are nodes (or points) and threads are the lines or edges of the graph. Each button is also a *binary variable* in that it is either connected (tied) to another button or buttons, or it is not. Thus the toy model provides us with a binary idealization of a network of primitive relations. Moreover, the buttons as relations play a *triple* role: each button as relation serves as an *ingredient* in a relation relating terms, or as a *product* of a relation, or as a *catalyst* for yet another reaction. With a reaction, yet another relation relating terms is either formed or broken. A relation becomes a term in yet another relation, producing relations of relations of relations. . increasing their diversity.

An essential feature of random graphs is that they show regular statistical behavior as one "tunes" the ratio of threads to buttons. As Kauffman notes,[26] a *phase transition* occurs when the ratio of threads to buttons passes 0.5. At that point, a cluster forms. Even in small systems, with only 20 buttons (relations), a cluster forms when the ratio of threads to buttons is half, that is 10 threads (connections). With larger systems, say 10,000 buttons, a large cluster or component of the system would emerge with about 5,000 threads. With an increase in the ratio of threads to buttons, more isolated buttons and small clusters become cross-connected into the larger cluster. As applied to primitive relations, what we find when we pick up any given relation is that it is attached to a larger interconnected cluster of many more relations.

Graphically represented, with the *y*-axis the size of cluster and *x*-axis the thread-to-button ratio, we find a sigmoidal curve that rises steeply once the ratio of edges (threads) to nodes (buttons) passes 0.5. In sum, we have an obvious *nonlinear* function. Applied to the dynamics of knowing, what we get when the relations are increasingly connected together are relations of relations of relations, just as Russell, and James before him, had envisioned. That is, computationally, the *length*[27] of relations increases. With random graphs, we also get a formal tool exhibiting regular statistical properties to address higher order emergent properties of clusters of relations as they are formed, that is as the density of connections in a graph of relations become greater. I will address these properties in greater detail shortly.

But there is an important limitation on the button and thread toy model which should be explored because it is ultimately tied to what is called the binding problem. This problem is an explanation of the *unity of consciousness.* We can say that it is the problem of explaining the unity of primitive relations of immediate awareness. As Kauffman notes on his own use of random graphs, the act of tying or connecting buttons does not in itself create still more buttons and still more threads. Buttons and threads cannot do this. But epistemic primitive relations, such as the primitive somatosensory-motor relations and the primitive relation of imagining that act on or relate other relations of our knowing, do create other epistemic "products" or relations of knowing. There is a real sense as noted above in which as clusters of relations are formed, they act as *catalysts* for the formation of yet more relations. In the case of kinds of knowing found in Boundary Set *S*, they create the *manner* by which one exhibits *knowing how*.

When someone knows how to do something, whether it is as apparently as simple as tying one's shoes, driving and parking one's car, knowing when and with how much pressure to "put on the brakes," or as complex as solving quadratic equations, there is form and structure to their doing. There is a manner (defined in part by timing) which is not found in the doing of one who does *not* know how. The clusters of primitive epistemic relations of *knowing the unique* and patterns of action of *knowing how* are formed giving rise the seamless and smooth timing and thought found in the doing of one who knows how.

To some extent the binding problem here can be interpreted as the problem of explaining how the *manner* of *knowing how* arises as a set of emergent properties of Boundary Set *S*. On one level, analysis of how emergent properties of knowing arise or are catalyzed to form the seamless and smooth manner of *knowing how,* was given earlier in my examination of the primitive relation of *imagining* and its further relation to the hierarchy of the senses and the primitive relations of *moving* and *touching* in the performance of a surgeon probing an incision or wound. Indeed, an understanding of the cluster of interrelations [relations of relations] among these primitive epistemic relations making up immediate awareness is necessary to understand *knowing how*. And one cannot obtain an understanding of these relations by inspecting the nervous system. It must be gotten from analysis of the relation(s) of knowing itself, between a subject, S, and an object(s), O, and the kinds of interactions and transactions which can occur among them.

5.10. Autocatalysis of Knowing: Some Law-like Properties of Immediate Awareness and the Binding Problem: Rule-Boundedness

As noted above, what we get with random Boolean networks is a theoretical means by which to understand kinds of knowing in Boundary Set *S*. It gives us a means to understand immediate awareness incremental *learning* (or *coming to know*), provided we have an adequate classification of primitive relations themselves, their relations to one another, and their relations to patterns of *knowing how*. Applied to Boundary Set *S*, this is the problem Polanyi recognized when one makes something that may be distal nonetheless function as a proximal term in an immediate relation of doing. It is also the "binding problem" originally stated by Russell as follows: "What I demand is an account of that principle of [primitive] selection which, to a given person at a given moment, makes one object, one subject and one time intimate and near and immediate, as no other object or subject or time can be to that subject at that time, though the same intimacy and nearness and immediacy will belong to these others in relation to other subjects and other times."[28]

Alternatively, the binding problem has been described by Koch and Crick [1990] as it pertains to the visual system, as follows:[29]

> We suggest that one of the functions of consciousness is to present the result of various underlying computations and that this involves an attentional mechanism that temporarily binds the neurons together by synchronizing their spikes in 40 Hz oscillations. These

oscillations do not themselves encode additional information, except in so far as they join together some of the existing information into a coherent percept.

But Crick and Koch's *physical* theory will not answer Russell's demand. Even where we have a complete physical theory of relevant neuronal activity, we are still left with the question. Why does *this* physical process give rise to *this* experience of immediate awareness? A still further alternative forming of the binding problem is as follows: What controls how terms of specific primitive epistemic relations are combined? Moreover, given the immense if not uncountable number of primitive components involved, what combines them and how are they combined? Concerning the visual system alone, the problem is enormous. As explained by Scott:[30]

> . . .there are an "almost infinite" number of visual patterns (I would say an immense number) that can be recognized, so it is not possible to assume a "grandmother cell" that corresponds to each pattern. Thus recognition must be related to the activity of a *set* of neurons, but this leads immediately to the *binding problem* because recognition of a single pattern must involve neurons in several different visual areas. In order to be bound together. . .the participating neurons must carry a common label, and they suggest that electrochemical oscillations in the 40-70 hertz range bind the relevant neurons in short-term memory. . .an idea that goes back to William James (1890).

On another level, however, the use of random graphs provides us with a more detailed analysis of the properties of the interrelations among the relations. I have used the word 'catalyst' above to refer to a primitive relation which acts to form yet other primitive knowing relations. The structure and dynamics of the process of what chemists call *catalysis* can be used to understand the dynamic self-organizing nature of immediate awareness as a set of primitive relations in relation to *knowing how*.

Briefly, we may understand catalysis of knowing in Boundary Set *S* in the following way: Some primitive relation A (one of our buttons in our use of Kauffman's toy model) might combine with relation B (button) to make C. However, there might be another primitive relation D, a catalyst, which causes A to combine with B very quickly. *Knowing how* to solve certain problems, say in mathematics (but also chemistry and other disciplines), sometimes depends upon having an image in one's mind of the *form* of the solution, as Kekulé's benzene ring, but without the equations which are the public knowledge *that* content of the solution. We may say that the closed ring image of a snake biting its tail is D, a primitive epistemic object of the primitive relation of imagining, which acted as a catalyst for Friederich Kekulé, like the slots of a lock in which the keys, the relation A [knowing the form of carbon atoms] and the relation B [knowing the form of hydrogen atoms] fit into to make C, the benzene ring solution.[31] In sum, D was the catalyst joining A and B to make C. Any of these then may (and have) become catalysts for other knowing relations.

Catalysis of this kind occurs because there is an intermediate state between two (or more) relations A and B in which one or more terms among the relations are distorted. In the dynamics of the change of interrelations and transformation of the relations, the direction or "sense" of the terms may be unclear. That is, for example, just as in a chemical reaction when certain molecules (relations) are joined, certain of their atoms (terms) may be affected and others not. Of those affected, there is a sharing, saturating, or "smearing" of the atoms and their electrons such that the molecules are bonded together. But the sharing or saturating is never complete, otherwise at least certain of the atoms would be destroyed. Likewise, if one examines the primitive somatosensory *cum* motor relations of even relatively simple tasks, one finds a like bonding and sharing or "smearing" of terms among the cluster of relations involved.[32] This is a sharing of particulars such as images, felt textures and space of objects of touch as one moves, and likewise a sharing of particulars of smells, colors, and other terms of the senses, including those distal terms that are functioning as proximal. In a sense, the object(s) of immediate awareness (of *knowing the unique*), the primitive relations of knowing, are *incomplete* as are the indexically functioning names we sometimes use to refer to them, along lines originally recognized by Frege.[33]

In essence, there are two parameters involved in our use of Kauffman's toy random graph to characterize primitive epistemic relations. These parameters are the *diversity* of primitive relations (Rp^d) and the *probability* (Pr) that any given relation will catalyze another reaction. A "reaction" on a relation here might be understood as a transformation or change of a relation by entering or combining with other relations or ending a relation. Because random graphs show regular statistical behavior, we know certain core phenomena involved in networks of primitive relations, stated in (1) and (2) below. These are some law-like properties of networks of knowing. But we are also left with many unknowns, such as the crucial one stated in the form of a question (3):

(1) If we increase the number of connections among the primitive relations, then we increase the diversity of primitive relations of knowing found in the network and the sharing or "smearing" of their terms.

The proof for (1) has already been established above. But basically what happens is that as the number of connections between the relations increases, each relation which is connected becomes a candidate to catalyze reactions by which new relations are formed. Moreover, we get the following:

(2) As the diversity of relations increases, the ratio of connections to relations also increases.

That is, as the number of lines connecting relations increases, the diversity of relations in the system increases, as stated in (1); then those relations themselves become candidates to further catalyze still further reactions by which the relations themselves are formed. Though all this is on analogy with chemical reactions, these are general properties found in random networks where the ratio of elements or relations to connections among the elements exceeds a certain level. Moreover, it accords well with what we know empirically of the phenomenon of machine learning (and coming to know) in actual neural networks,[34] including those that characterize the somatosensory *cum* motor system. As our knowing increases, the means by which we come to know even more increases (to a certain threshold).

Another of the core phenomena involved in the random graph of primitive epistemic relations which we want to understand is the following:

(3) What is the probability that a given primitive relation would catalyze another reaction [of a relation]?

An answer to (3) would tell us what set of primitive relations might be collectively self-sustaining or (to use another chemical term) *autocatalytic*, or self-catalyzing, thus giving rise to immediate awareness of an object. Of course this is the kind of information we cannot have short of considerable empirical research, particularly on the somatosensory-motor system. Though such information may be sparse, there is currently much research underway related to neural activity onset to exploratory moving and touching.[35] Nonetheless, short of that information, an understanding of the emergence of self-organizing and self-sustaining networks, that is collective autocatalysis, is possible to some extent with an understanding of the nature of reaction networks which are themselves subgraphs of the larger random graph of relations. A subgraph is a restriction on a relation (or set of relations).

Earlier, I noted the usefulness of digraph theory as a formal means to become clear on the fundamental properties of Boundary Set *S*. Where digraphs, such as Kauffman's random graphs, are presented as path diagrams meeting the requirements for path analysis [that is where the connections are asymmetrical, as they are in the present example], the density and connectedness of the digraph indicate whether connections are missing. Density is a property of such graphs which can be quantified: it is the number of direct connections over the number of possible connections, given by the following equation stated earlier: $D = DC/N(N\text{-}1)$. Where 'D' stands for density; 'DC' stands for number of direct connections; and '*N*' stands for the number of properties.

Obviously, density cannot fall below some minimal value because less than *N*-1 direct connections results in some properties not being connected. On the other hand, as Kauffman's research has shown, if density of connections is too great, a network will veer into chaos. There are two *P* parameters which can be modified so as to "tune" networks with $K>2$ so that they are orderly: density and sparseness of connections. Tweaking the *P* parameter can make a chaotic network become orderly.[36] As he states, if different networks are built with increasing *P* biases, starting from a no-bias value of 0.5 or slightly greater than 0.5 to the maximum value of 1.0, networks with $P = 0.5$ or only slightly greater than 0.5 are chaotic. Networks with *P* near 1.0 are orderly. The research cited by Kauffman shows that there is a critical value of P where a network will switch from chaotic to ordered. That critical value is the edge of chaos. And it is important to point out, as Kauffman does, that these rules apply to networks *of any kind*. This includes networks of primitive relations of immediate awareness, and, as noted earlier, I believe that kinds of knowing found in Boundary Set *S* are poised right on that edge of chaos with that critical value *P*. Knowing systems at the phase transition between order and chaos are those best able to exhibit ordered but flexible behavior in their knowing how, the kind we saw above in a surgeon's and tennis player's *knowing how* and immediate awareness.[37]

Moreover, the above noted regular statistical properties of random graphs also shows that even if we assume that a given relation has a fixed chance of being able to function to catalyze a given reaction, to the extent that the properties of random graphs apply to primitive knowing relations of somatosensory-motor systems, when the set of relations reaches a critical diversity, a component of catalyzed reactions emerges.

5.11. A Random Boolean Network of Knowing: The Emergence of Order

To understand more clearly how reactions among primitive relations of immediate awareness and relations of *knowing how* give rise to the seamless, smooth oscillations and timing found in *knowing how* behavior, we can present an even more useful idealization of the above networks. In principle, each primitive relation can be turned off or turned on by other relations in a reaction network. That is, it can serve as an ingredient of a reaction, a product, or a catalyst for another reaction. But unless there is a source for stability or order the potential for unrestrained chaos is evident. As is evident to some degree with the regular statistical properties on these networks above, the order emerges from the collective dynamics of the networks of relations themselves.

To see how order emerges, we can idealize the above graphs of relations as Kauffman uses random Boolean networks. These are massively parallel, highly distributed processing systems of binary variables or nodes, such as the buttons or relations in the above toy model, each with two possible states: *on* and *off*. These variables are coupled to one another such that the activity of each element is governed by the prior activity of some of those elements according to a Boolean switching function. That is, the state of each variable (primitive relation) or node at time $t+1$ is determined by its own Boolean function, which takes as input the states of other variables in the network where this is indicated by a directed line between the variables at time t.

We can again think of a highly distributed and massively parallel network or cluster of networks of primitive knowing relations connected by lines to illustrate their emergent properties. Idealizing the behavior of each primitive relation as a simple binary variable [an *on* or *off* variable], we can imagine each primitive relation as a switch which turns on other relations, as in the Kekulé example, or turns others off. For example, the primitive relation of imagining a certain internal surface to a wound can turn on the relation of touching and relation of moving with a probe in *anticipation* of that internal surface. The image of the internal surface is a term not only of the relation of imagining but also of the relation of touching and moving, related by imagining. Thus we have a relation relating another relation, with sharing of terms.

In any case, the essential point is that some relations can turn others on, while others can turn yet others off. If the relations are initially tied together randomly, we find emerging order. Under certain conditions, the randomly switched on and off relations settle into coherent, repeating patterns.

A vast network of such relations tied together can take on a large number of possible states of knowing, its *state space* (Poincaré map). As noted above, this is the mathematical universe of possibilities in which a system is free to roam. Among those possibilities, either all the relations might be on, or they might all be off, or in between these two possibilities they can assume a vast number of possible state combinations. If we have a knowing system consisting of 20 primitive relations, each of which can be in one of two possible states, *on* (represented usually with 1) or *off* (represented with 0), the number of possible configurations is 2^{20}. If we have a system of 100, the number of states possible is 2^{100}. Where we have multiple levels of relations of relations, a knowing system consisting of at least 100 relations is not very large.

In our random network of primitive relations, each relation is regulated by others that serve as inputs. The dynamic behavior of each relation, whether it is on or off at the next moment, is governed by a Boolean function. The activity of each primitive relation is in response to all the possible combinations of activities in the input relations. The Boolean function, OR, indicates that a relation is active if *any* of its input variables is active or on. The AND function indicates that a relation will become active only if *all* its inputs are currently active. Thus it is possible to calculate the number of functions which could apply to any binary element (relation) in a network. For example, if a binary element has K inputs, then there are 2^K possible combinations of inputs. Either an active or inactive result must be indicated for each combination. That is, there are 2 to 2^K power possible Boolean switching rules for that element.

We can formalize what has been said thus far by noting that random Boolean systems contain N elements which are linked by K inputs per element. Inputs and one of the possible Boolean functions are assigned at random to each element. By assigning values to N and K, one defines a class of networks with the same local features. For example, with a simple network consisting of 3 relations, then we have 2^3 possible combinations. With each combination of binary element [that is relation] activities constituting one network state, all the elements (relations) of each state assess the values of their regulatory inputs at a given moment in time. The succession of network states is the *trajectory* of the network, that is the *trajectory of knowing*.

Because there is a finite number of states of the network, any Boolean network will reenter a state that it has previously encountered. It will cycle through the same states it has cycled through previously. The sequences of states that the network flows through are its trajectories, and more than one trajectory can flow through the same *state cycle.* State cycles are the *dynamic attractors* of the network, and once the trajectory of the network carries it onto a state cycle, it will tend to remain there.[38] Thus, as Kauffman points out, Boolean networks settle down to a state cycle whether the number of states is small or astronomically high. If a system falls into a small state cycle, it will behave in an orderly manner, but if the state cycle is too vast, the system will behave in a manner that is essentially unpredictable.[39] The collection of trajectories flowing into a state cycle or that lie on it constitutes the *basin of attraction*[40] of the state cycle, and every network *must have at least one state cycle* (and it may have more than one). The *length* of a state cycle is the number of states on the cycle, ranging from 1 for a *steady state* to 2^N.

Attractors are the source of order in large complex dynamical systems. They are of the essence in Boundary Set *S*. In the mathematical Poincaré space of possible behaviors, it is the attractors which trap the dynamics of the system into subregions [subgraphs] of its state space. But this is not sufficient to ensure orderly dynamics. There must also be conditions or properties of the networks which resist perturbations. As Kauffman points out, there are ways of "tuning" networks where $K>2$ so that they are orderly and not chaotic. The requirements for order are not in how we construct our primitive relations together, but only that they be sparsely connected.

With the interpretation of attractors in knowing systems as the primitive relations of immediate awareness and patterns of action of *knowing how*, the attractors become the sources of the order we observe exhibited by one who knows how to do those tasks or performances found in Boundary Set *S*. The enormous number of "doings" of a person who knows how to do something will follow trajectories across a space of possible doings that will flow into attractors, families of trajectories. That is, the knowing system, a person doing something they know how to do, will settle into an orderly few doings. No matter what initial point a dynamical system starts from, it ends up moving around on some well-defined shape in phase [Poincaré] space.

But that person who knows how to do something, say performing surgical tasks, must also be resistant to disturbances or perturbations. As with any living organism, if a natural knowing system is disturbed say, with doubt, its trajectory through time may very well change. Disturbances can originate internally to a system or externally. In fact, we may say that since knowing is open-ended and natural knowing systems are not complete or "all knowing," they must *necessarily* experience perturbations or disturbances as long as they are alive, and most certainly as long as they are actively inquiring systems, not merely reactive.[41]

5.12. The Boundary of Epistemic Boundary Set *S*

It is important to realize that lines traced in a phase space are usually oscillations, that is some form of *sigmoidal* function. Because the trajectory traced by *knowing how* is sigmoidal, the *knowing how* requires the "suppression of noise," the suppression of any point or part of the line tracing outside the set of knowing into the ground or complement of the object or goal [or pattern of action] of the knowing. Recall our earlier discussion of object or figure and ground and how the ground functions as an index to the object of knowing. The *manner of knowing how* is the *figure* [object] and all else is the *ground* [noise]. There is a sense in which the ground, *points lying outside the set of knowing in the complement of that set,* function indexically and also limit the range of the dynamics of the knowing behavior.

Even if we attempted to mathematically characterize just the physical performance found in this set with even an unlimited set of differential equations, such an effort would prove intractable and fail even though we are addressing a deterministic system. In part, it would fail because we would still not be able to precisely measure and specify the initial conditions characterizing the state of the knowing at time t so as to do mathematical calculations determining the knowing at time $t + 1, t + 2, \ldots t + n$. The most precise measurements yet made in physics are accurate to only about nine decimal places,[42] and for these kinds of dynamic performances accuracy to nine decimal places (or to any finite number of decimal places) is useless for prediction.

Very small changes in initial conditions, for example *even the angle and distance at which a surgeon initially perceives an incision,* can lead to enormous changes in the actual performance of a probe of that incision. Moreover, the number of calculations necessary to characterize the dynamics of the performance would be overwhelming. The phase space portrait of even a single performance of surgical probing involves an immense number of epistemic elements in combination, making it impossible to visualize graphically or reduce to a finite set of differential [or other] equations. In sum, *knowing how* embedded with immediate awareness, our *knowing the unique,* is a complex dynamic system. Even though the behavior exhibiting the knowing may be deterministic, it is not predictable, and no two performances of surgical probing are alike.

If we were to record a surgeon performing such a probing task by film, digitizing our results, we would be left with a series of line tracings, providing a two-dimensional copy of the doing. Plotted on a diagram, the multidimensional *knowing how* with embedded immediate awareness hyperphase space could be reduced to two-dimensional trajectories. That is, though the state space itself may be high dimensional, actual flow of trajectories may be restricted to some low dimension. Each motion of a surgeon's hands and fingers, smoothly performing a digital insertion while palpating the surrounding area of a victim's body with the other hand, would show up as lines with varying oscillations "smoothed" out by noise suppression with a limited dynamic range. Close examination of even small regions of any part of a given line or trajectory of knowing, that is the boundary of the knowing making up epistemic set *S*, however, would reveal that it is not sharp but "fuzzes" out into puff layers.

The boundary of our epistemic set *S* is *like* what mathematicians refer to as a *chaos fractal basin boundary.* As noted earlier, I am here interpreting dynamic alternative attractors as epistemological categories previously sorted, focusing upon *knowing how.*[43] The set of phase state cycles are the dynamical attractors of a knowing network, and once the trajectory of the network carries it onto a state cycle it will tend to remain there. This set of states that flow into a cycle or that lie on it constitutes the *basin of attraction* of the state cycle. A digitized film of the performance will show that there are competing types of motion, that is attractors, found in the performance and each motion has a basin of initial states producing it. However, boundaries between these basins do not form a precise line, but amount to a fuzzy fractal edge. This is why *knowing how* to probe a wound is *like* a chaos fractal basin boundary. The most famous instance of such a mathematical object is the Mandelbrot set and lesser known Julia set. The use of such characterizations is appropriate not only because of the infinite number of possible primitive relations and terms affecting the somatosensory-motor network making up immediate awareness, but also because of the complex dynamics making up patterns of action of *knowing how*, as well as the complex dynamic interaction among both these sets.

Moreover, most computational problems arising in robotics and neuroscience, directly germane to mathematical characterizations of *knowing how* control systems, and computational geometry, have as natural domains the reals or complex numbers.[44] Each epistemic set must be mathematically defined over the reals and complex numbers as is the Mandelbrot and Julia sets, generating certain of the same kinds of complex dynamics.

One cannot know from an inspection of the mathematical equations generating such complex dynamics what kind of attractor will in fact be generated from solutions of those equations. One must simply input the equations into a computer and *wait to see what the computer does.* Again, one cannot obtain *knowing how* with embedded immediate awareness, our *knowing the unique,* from *knowledge that.* There is a real sense in which that is precisely what medical schools realize in the training of surgeons: They input much *knowledge that* but emphasize much more first-hand familiarity and direct experience to cultivate *knowing how* with immediate awareness in surgery students and interns. When the appropriate need arises, they wait to see what the surgeon-trainee will do.

5.13. Parameter Space and Rugged Landscape of Boundary Set *S*

Obviously the behavior generated in our Boundary Set *S* depends upon a number of parametric values such as the timing with which touching and moving proceeds as well as how rapidly the terms of say, the primitive sensory and somatosensory network relations are added or removed. We might artificially hold these values as constants though in fact they also change, sometimes gradually and sometimes rapidly. The point here is that as these values change the trajectories and the attractors also change in their flow across the phase space of possible trajectories of the knowing behavior. On a more macroscopic level, we can think of the structure, that is parameters of the landscape or space of knowing possibilities underlying the process of coming to know or knowing. We could then evaluate the *fitness* of a given natural knower as that knower traverses the space of possibilities. We might think of that structure as fixed, while in reality the structure actually changes as a result of interactions and transactions with the environment and with other knowers, as well as changes in that environment and in others. If we hold the parameters constant, we could characterize the flow across the space of possibilities as an adaptive walk or search across the fitness surface, where the peaks are sought. If the parameters change, the flow is far more complicated, and it is precisely under conditions of changing parameters that I must view epistemic Boundary Set *S*. The variability of knowing behavior of Boundary Set *S*, as the structure or parameters of a system, are altered can be characterized as the ruggedness of a fitness landscape. I am referencing the use of the *hill-climbing* framework used by Kauffman,[45] borrowed from Wright.[46]

5.14. Summary

An important facet of the analysis I gave earlier showed that manner exhibiting *knowing how* is not only an emergent property of complex interrelations among the primitive epistemic relations of the senses, imagining, touching and moving, but also of complex interrelations among their differing spatial (as well as temporal) relations with the body. Not only do the senses, imagining, touching and moving form an epistemic hierarchy with touching, imagining, and moving on a more complex level than any of the senses, *their spatial relations with the body* are also part of that hierarchy. One who knows how exhibits a seamless, smooth and refined manner of doing not found in one who does not know how. That knowing includes anticipatory imagining of structures and patterns of doing. The seamless, smooth character or manner of a doing with one's body is a product of that complex cluster of multileveled primitive epistemic relations of relations of the senses, imagining, moving, and touching, as well as a use of their spatial relations to the body.

Moving is treated as a complex sensory *and* somatosensory-motor phenomenon in the neurophysiological literature,[47] but it can also be intentional. The concept *touching* is clearly *bodily* intentional, in the sense that we use our bodies cognitively and indexically when we touch, whereas tactile feeling is not intentional.[48] As noted, the space of feeling and the space of touching are not identical because of the relation of imagining, including anticipatory imagining to touching. Again, touching requires that one intentionally heed and focus upon the object of touching with one's body, whereas one can experience with one's senses [*e.g.* tactile feeling] without that kind of intentionality.

Moreover, this intentional heeding and focusing will differ epistemically in its structure depending upon whether or not one has visual access to the object(s) or whether or not one is touching the object directly with one's body [for example with a finger], or if the touching is mediated by an instrument used as a probe.[49] It is significant to point out that we use all kinds of instruments to probe, from physical probes such as surgical instruments, to *abstract concepts*, including varying abstract spatial structures. With the latter, though the primitive physical somatosensory *cum* motor relation of touching may not be involved, the primitive relations of imagining and moving are. One can know how to probe an incision, a dark environment, or unexplored abstract structures of mathematics. Whether probing an incision with a surgical instrument or probing abstract structures in hyperspace, one continuously forms images of the object(s) of the probing. Those objects may be the particulars making up the configuration of the inside of the wound, (as well as the item that produced it), or they may be the structural configurations of kinds of space.

Again, the primitive relations of moving and touching are not identical. Moving which has epistemic significance is clearly intentional and requires a focal heeding with one's body, or kinds of spatial relations with it, as does touching. These relations are made even more complex because of the relation of imagining [and especially imagining which is anticipatory] to moving, and also because touching is more a *close*[50] relation than moving which can be more distant. With touching, the body is clearly used indexically in a very close, concrete way with proximal objects. With moving, the indexical function of the body may be more abstract because it can involve objects (including imagined and anticipated patterns) which are at great distances from the body or are abstract structures. Dancing is another example of moving, in addition to surgical tasks, in which one anticipates imagined patterns of movement to come.

This chapter has also presented a more formal approach to the subject of immediate awareness as we find it in knowing how,

Boundary Set *S*. I have set forth the theoretical framework and approach, using set theory to sort the major categories of both the epistemic universe and the epistemological universe. I have also more precisely defined Boundary Set *S*, and set up the mathematical framework within which Boundary Set *S* must be approached. A number of information-theoretic and other measures were also included to show how the self-organizing and adaptive network of primitive relations in Boundary Set *S* could be measured.

I used Kauffman's random Boolean networks to show how they might be used to understand the dynamics of our Boundary Set *S*. Since the properties of Boolean networks have been thoroughly studied and techniques have been developed for determining the dynamical properties of specific kinds of networks,[51] they are of tremendous value in studying the fundamental properties of highly complex dynamical systems which may have very large, even immense, numbers of interrelated components.

Given their core properties, switching Boolean networks are important for an adequate theory of *complex* but *ordered* systems, such as the interacting neurons in a neural network or interacting primitive relations in a natural intelligence network. This is so because these networks facilitate handling the large number of elements involved as well as their interrelations. They are also important because one can trace the behavior of populations of very large numbers of interacting elements making up different network configurations [with varying parameters] across landscapes [Poincaré maps or spaces of possibilities], so as to evaluate their properties.

---

[1]My analysis of this task is not intended as definitive of the steps for doing it. Those steps differ relative to wound presentation, overall medical condition of the victim, and conditions for conducting the task, including the availability of surgical tools, medications, including anaesthesia and other aids. For example, how the task is performed can differ depending upon whether one is in a controlled sterile environment such as a hospital surgical OR, or uncontrolled environment such as in combat. I am assuming the most primitive conditions.

[2] This was especially the case in combat conditions that existed in prior wars fought by the U.S. Those kinds of conditions no longer generally exist.

[3]The terms 'proximal' and 'distal' are borrowed from anatomy, but can be used to unfold the structure [or anatomy] of our knowing.

[4] See Stephen Kosslyn, "*Visual Mental Images in the Brain: How Low Do They Go*," presented at a meeting of the American Association for the Advancement of Science on the Cognitive Neuroscience of Mental Imagery, February, 2002. According to Kosslyn, using images this

way also causes the same effects on memory and the body as occur during actual perception, but the two functions are not identical.

[5] The description here assumes few technological assists such as X-ray, as in severe combat conditions.

[6] Gary Stix, "Boot Camp for Surgeons," in *Scientific American*, September 1995, p. 24.

[7] Ian Stewart, *Nature's Numbers*, New York, Basic Books, 1995, p. 123.

[8] Throughout this section, I follow Stuart Kauffman's use of random Boolean networks. See his *The Origins of Order: Self-Organization and Selection in Evolution*, Oxford University Press, 1993, and his *At Home in the Universe: The Search for the Laws of Self-Organization and Complexity*, Oxford University Press, 1995.

[9] Ian Stewart, *Nature's Numbers*, New York, Basic Books, 1995, p. 100.

[10] M. Estep, "Toward Alternative Methods in Systems Analysis: The Case of Qualitative Knowing," in *Cybernetics and Systems Research, Vol. 2,* Robert Trappl, (ed.), Elsevier Science Publishers B.V. (North-Holland), 1984.

[11] I am using the term *immense* as defined by Walter M. Elsasser, *Atom and Organism: A New Approach to Theoretical Biology*, Princeton University Press, 1966, as cited by Alwyn Scott in *Stairway to the Mind*, Springer-Verlag, 1995. An immense number $\mathfrak{I} = 10^{110}$ In contrast to a finite number of items which can be put on a list and examined, for an immense number of items this is not possible. There would not be sufficient memory capacity in any computer which could ever be built to store an immense number of items.

[12] E. Steiner, *Methodology of Theory Building*, Sydney, Australia, Educology Research Associates, 1988.

[13] F. Harary, *Graph Theory*, Massachusetts, Addison-Wesley, 1969, p. 199.

[14] M. Estep, "Toward a SIGGS Characterization of Epistemic Properties of Educational Design," in *Applied General Systems Research*, George Klir, (ed.), NATO Conference Series, New York, Plenum Press, 1978, pp. 917-935.

[15] See Ralph Grimaldi, *Discrete and Combinatorial Mathematics*, Third Edition, Reading, Massachusetts, Addison-Wesley Publishing Company, 1994, pp. 374-375.

[16] For the sake of argument, we assume the epistemic universe is like the *filled* Julia set of a polynomial map on a Riemann sphere where $S=C \cup \{\infty\}$ of the form $g(z) = z^2 + c$. The boundary Julia set is the set of points that don't go off to infinity under iterations of g. [See Blum, 1989]. Boundary Set *S* is rule-bound in this sense.

[17] James Albus, *Brains, Behavior, and Robotics*, Peterborough, New Hampshire, BYTE Books, 1981.

[18] Hartley, 1928.

[19] I arbitrarily sorted these kinds into a matrix of one/many paths and one/many termini. 'Pr' stands for "protocolic," which means one path to one terminus; it is also possible to have a performance with one path but leading to many termini; 'Co' stands for "conventional," which means many paths leading to a single terminus or many paths leading to many termini. 'In' stands for "innovative," which means combining one or more given paths with one or more given termini in new ways; 'Cr' stands for "creative" which means producing new paths or new termini.

[20] Ian Stewart, *Nature's Numbers*, New York, Basic Books, 1995, p. 117.

[21] Ibid., p. 94.

[22] James L. McClelland and David E. Rummelhart, *Parallel Distributed Processing, Volumes 1 and 2,* Cambridge: MIT Press, 1986.

[23] Stuart Kauffman, *At Home in the Universe: The Search for the Laws of Self-Organization and Complexity*, New York, Oxford University Press, 1995, p. 18.

[24] Ibid.

[25] Bertrand Russell, 1984, p. 80.

[26]Kauffman, 1995, p. 56.

[27]Computationally, the size of a problem instance is the *length* of (the description of) the instance, measured in some standard units, usually binary.

[28]See Russell, 1984.

[29] Alwyn Scott, *Stairway to the Mind*, Springer-Verlag, 1995, p. 129f, gives an excellent summary of Koch and Crick's statement of the problem.

[30]Ibid., p. 129.

[31]See Cohen and Stewart, *The Collapse of Chaos*, New York, Penguin Books, 1994, p. 41. I have obviously used an example which is not strictly bodily kinaesthetic, but the relation between primitive relations such as imagining and other kinds of knowing is evident.

[32]I believe that this is the kind of image of function as "unsaturated" that Frege had in mind when he wrote of *incomplete names* in his *Über Begriff und Gegenstand, Der Gedanke*, and *Gedank engefüge*. Though it generally is not recognized, Frege wrote of two kinds of names, the logically proper name which has both sense and denotation [and is, thus, a proper object of propositions or thought], but also incomplete names which are functions (and concepts). The former logically proper names are those ordinary names appearing in our language and have objects, that is, they are "saturated." The latter incomplete names, on the other hand, do not strictly speaking, have objects, according to Frege, because they are unsaturated. I believe what Frege meant by "unsaturated" is [by analogy] what a chemist means by the "smearing" or "sharing" of electrons of atoms of molecules. The "sharing" or "saturating" is never complete. By analogy, the object(s) of immediate awareness, the *sui generis* objects of the indexical function of knowing the unique, are likewise unsaturated or incomplete because of the incompleteness of the primitive relations of a Subject to them. In certain senses, Frege had a clearer conception of the object of [proper name in the indexical or individuating sense] knowing by acquaintance than did Russell.

[33]See Gottlob Frege, "Über Begriff und Gegenstand" in *Vierteljahrsschrift für wissenschaftliche Philosophie*, 16, 1892, pp. 192-205; and "Der Gedanke" in *Beiträge zur Philosophie des deutschen Idealismus, 1,* 1918-1919, pp. 58-77*;* and "Gedank engefüge" in *Beiträge zur Philosophie des deutschen Idealismus, 3,* 1923-1926, pp. 36-51.

[34]I am referring here to Hebb's postulate of learning. See D. O. Hebb, *The Organization of Behavior*, New York, Wiley, 1949. According to Haykin, [Simon Haykin, *Neural Networks: A Comprehensive Foundation*, New York, Macmillan, 1994, pp. 51-53], Hebb's postulate has been a subject of intense experimental interest among neurophysiologists and neuropsychologists for many years. Empirical research has shown that "a time-dependent, highly local, and strongly interactive mechanism is responsible for one form of long-term potentiation (LTP) in the hippocampus." LTP is a use-dependent and long-lasting increase in synaptic strength that may be induced by short periods of high-frequency activation of excitatory synapses in the hippocampus. Experimentation showing that LTP in certain hippocampal synapses is Hebbian appears to have been replicated by other investigators.

[35]See Miguel A. L. Nicolelis, *et al.*, "Sensorimotor Encoding by Synchronous Neural Ensemble Activity at Multiple Levels of the Somatosensory System," in *Science,* AAAS, Vol. 268, 2 June 1995.

[36]Kauffman, 1995, p. 84, cites the research of Derrida and Weisbuch showing this.

[37]The relations between these kinds of ordered but flexible behavior in humans and lower animals have been explored to some degree in research into the rhythmic movements of the rat's trigeminal system. That system is a multilevel, recurrently interconnected neural network generating complex emergent dynamic patterns. See Nicolelis, *et al.*, "Sensorimotor Encoding by Synchronous Neural Ensemble Activity at Multiple Levels of the Somatosensory System," in *Science*, AAAS, Vol. 268. 2 June 1995.

[38] Kauffman, 1991. Also see Stein, 1988.

[39] Kauffman, 1995, p. 78.

[40]The concept *basin of attraction* is given meaning within stability theory. 'Stability' is a general term characterizing response to perturbations. Stability applies to different aspects of a dynamical system and can refer to individual points (local stability), to trajectories (asymptotic local stability), to families of trajectories (attractors), or to an entire dynamical system (that is, whether or not there is a unique attractor). Following Eubank and Farmer [1989], an attractor is an invariant set of components (points) that "attracts"nearby states. The *basin of attraction* is the set of points that are attracted to it. Formally, Ω is an attractor if there is a neighborhood N about it such that $F^t(N) \to \Omega$ as $t \to \infty$ and Ω cannot be broken into pieces $\Omega_1, \Omega_2, \ldots$ such that $F^t(\Omega_1) \cap F^t(\Omega_2) = \emptyset$. The basin of an attractor is the set of points attracted to Ω, that is $\{x: \lim_{t\to\infty} F^t \subset \Omega\}$. There may be many attractors, each with its own distinct basin of attraction in any given dynamical system.

[41]Much of supervised and reinforcement learning theory as applied to neural networks is based upon a reactive stimulus-response model, not an anticipatory, proactive model. Elsewhere, I have argued that the concept 'learning' and the concept 'coming to know' are neither identical nor equivalent. It is the latter that these theorists aim for, yet it requires epistemological analysis, not psychological analysis as does 'learning'. Where I consider the property of self-organization in neural networks, I will of course be addressing only unsupervised learning [or coming to know]. From this, it is also evident that I will not be concerned with backpropagation.

[42] Cohen and Stewart, 1994.

[43]At this point, I want to stress again that a great deal of caution must attend the use of dynamical systems theory and phase space models of Boundary Set *S*. My effort is to use the best and most precise mathematical tools and models available to a scientific as well as philosophic study of the nature of this kind of knowing which in terms of its epistemic properties has much in common with what Dreyfus earlier referred to as commonsense know how and understanding of human beings.

[44]This has been established over the past several decades in both engineering and computational neuroscience. For example, see D. J. Bell, *Mathematics of Linear and Nonlinear Systems*, Oxford University Press, 1990; Feldman, A.G., *Biophysics 11*, 565, 1966; various publications by Hiroaki Gomi and Mitsuo Kawato, most recently their "Equilibrium-Point Control Hypothesis Examined by Measured Arm Stiffness During Multijoint Movement," in *Science*, American Association for the Advancement of Science, Volume 272, 5 April 1996, pp. 117-120, and Blum, Lenore, *Lectures on a Theory of Computation and Complexity over the Reals (or an Arbitrary Ring),* Berkeley, International Computer Science Institute, 1989.

[45] Stuart Kauffman, 1993 and 1995.

[46] S. Wright, "Evolution in Mendelian Populations," *Genetics*, Vol. 16, number 97, 1931; and S. Wright, "The Roles of Mutation, Inbreeding, Crossbreeding and Selection in Evolution," *Proceedings of the Sixth International Congress in Genetics*, Vol. 1, number 356, 1932.

[47]See Berthoz and Israël.

[48]Substantial empirical research has established this claim, in addition to that of Berthoz and Israël. See Gardner's *Frames of Mind: The Theory of Multiple Intelligences,* Basic Books, 1993. See especially references included under bodily-kinaesthetic intelligence.

[49]A thorough analysis of the epistemic structure of *touching* requires an analysis of probes and their epistemic and spatial relations to our body. Moreover, what we know of the human use of the fingers to explore or come to know the texture and shape of objects has much in common with results of scientific neural experimentation with the rat trigeminal system. We know that rats rely on rhythmic movements of their facial whiskers *much as humans rely on*

coordinated movements of fingertips to explore or come to know objects in their proximal environment. The trigeminal system is a multilevel, recurrently interconnected neural network which generates complex emergent dynamic patterns of neural activity manifesting synchronous oscillations and even chaotic behavior [see Nicolelis, *et al.*, "Sensorimotor Encoding by Synchronous Neural Ensemble Activity at Multiple Levels of the Somatosensory System," in *Science,* AAAS, Vol. 268, 2 June 1995.

[50]Again, the terms 'close' and 'distant' as related to epistemic relations have meaning in relation to proximity with the human body, the ultimate instrument of all our external knowing. I am not happy with the distinction between touching and moving as I have left it here, and am not resigned to the distinctions between them as I have drawn them.

[51]See Stephanie Forrest and John H. Miller, "Emergent Behavior in Classifier Systems," in *Emergent Computation*, Stephanie Forrest, (ed.), Cambridge: MIT Press, 1991, pp. 213-227.

6. CAN NEURAL NETWORKS SIMULATE BOUNDARY SET *S*?

In this chapter, I will evaluate the extent to which neural network models can characterize the kinds of complex self-organizing dynamics found in Boundary Set *S*. In the last chapter, we saw that it is the attractors that become the sources of the order we observe exhibited by one who knows how to do tasks or performances found in Boundary Set *S*. But it is clear that attractors in neural networks will be chaotic unless there is some other ordering principle. Kauffman has suggested that principle may be learning, for example, Hebbian. However, there is a sense in which that cannot be so for kinds of knowing found in Boundary Set *S*. Among other things, those kinds of knowing are primitive relations between a subject and object where the latter is not a class object. Hebbian learning essentially proceeds by modifying synaptic weight so as to amplify changes throughout a network. Data is presented to a system input layer where the system algorithm operates upon it in terms of local rules, computing an input-output mapping with certain desirable properties. But all this is actually carried out on class objects in terms of some class rule, such as some rule of similarity based on Euclidean distance, defining a *class*. It is not carried out on unique, *sui generis* objects.

Setting the issue of *sui generis* objects aside, however, computer architectures differ immensely from one another. Top-down, logic based programming is a paradigm for knowledge representation systems, but is inadequate to address simulations of the kinds of behavior of interest here. Neural networks offer the best approaches to date, specifically in the form of self-organizing feature maps. Basically, a self-organizing feature map is one in which topographic maps are formed of the input patterns. The neurons are placed at nodes of a lattice and become selectively "tuned" to various input patterns (vectors) in the learning process. Over time, the neurons become ordered so that a meaningful coordinate system for different input features is created over the lattice. Thus the spatial locations or coordinates of the neurons in the lattice correspond to features of the input patterns.

Self-organizing feature maps are inspired by the way the human brain is organized. It is organized such that different sensory inputs are represented by topologically ordered maps on the brain. As we discussed in earlier chapters, the sensory and somatosensory-motor areas of the brain are mapped similar to layered "sheets" onto different areas of the brain. Our brain is organized relative to these different maps such that we are able to make associations between both spatial and temporal information that is streaming simultaneously, on multiple levels, to our sensory and somatosensory-motor systems. Our brains know how to reinforce connections between things that often appear together in those data streams, which makes it possible for us to be aware of our environment and to form higher order combinations of things. This in turn permits us to make sense of our experience and act intelligently in the world. The human brain is a natural, self-organizing feature map, building topographic maps onto itself.

Thus artificial self-organizing feature maps used in computers are designed specifically to be like the human brain in that respect. They are designed to be like natural brain "computational maps," sharing the same (or almost the same) functional properties. The computational maps are performed in Kohonen's Self-Organizing Feature Map (SOFM) by parallel processing arrays that can handle large amounts of information very quickly. They can rapidly sort and process complex input and represent the results in a simple and systematic form. We will specifically look at Kohonen's SOFM to see how it handles hierarchical streams of data and the formation of higher categories of things, to determine whether or not, and to what extent, it can handle kinds of knowing in Boundary Set *S*.

Additionally, the primitive relations of immediate awareness entail an indexical or "pointing" function. I earlier claimed that AI and AL approaches cannot handle indexically functioning proper names such as 'this', 'that', 'now', 'I'. Because indexicals exhibit a network and structure of *thought* contents when we use them to point or refer to items of experience as we experience them, they are highly context-dependent. These include both linguistic and non-linguistic indexicals, that is indicators occurring as words such as the above, or as gestures, images, and patterns of action. Since computers are classification machines of some kind, I want to expend some effort here to show why computers cannot handle even *language* indexicals let alone non-linguistic ones.

Of course, a prevailing view in the AI community (especially the "strong" AI community) is that everything is a computer. One needs solely to map things to sets of numbers and so long as we have appropriate algorithms to act on those numbers, we can, in principle, compute anything. In this chapter, while largely setting aside the issue of *sui generis* objects, I will look at various neural network architectures to see if they can handle the self-organizing and adaptive dynamics found in kinds of knowing in Boundary Set *S*. I will initially assume that at least in principle, some kind of neural network architecture will be able to simulate at least some of these kinds of knowing, and I will largely focus just on linguistic indexicals. I will then also look at some meaning representation languages to map linguistic indexicals into computers. My assessment of those efforts and my arguments will show that AI and AL approaches to linguistic indexicals fail precisely because those approaches conflate grammatical meaning with functions. 'Function' is taken to be a mapping in the mathematical sense as a set of ordered *n*-tuples. It will be useful here to thoroughly examine some very ordinary, concrete examples of the use of natural language context-dependent indexicals and their meaning to show this.

But first, we will take a look at a very ordinary situation in which normal human beings exhibit kinds of knowing how and immediate awareness found in Boundary Set S. Then we will discuss some basics of neural network architectures and what they can and cannot do.

6.1. The Cocktail Party Problem

Think of the last time you were at a party or gathering of some kind, or just in a crowded room with, say, five or more people. Everyone is milling around the room simultaneously in conversation on various topics, at various levels of interest (some dull; some highly animated and loud), at various decibel levels, and with music playing in the background as well. You may have been having a conversation with someone, but during your own conversation with the person in front of you, you nonetheless could overhear conversation of someone else in another part of the room. But you weren't paying any real attention to that other person and what they were saying or to whom they were saying it. Again, you really didn't pay any attention to them and what they were saying because you were engaged in your own conversation with someone standing right in front of you.

During your own conversation, however, midst all the other background sounds and conversational noise, the person you overheard in the background stopped talking. They stopped talking long enough for you to then take notice that they had stopped talking. You noticed because you no longer heard their voice in the midst of all the other noise and conversation, coming from the direction of the room where you heard their voice, even in the midst of your own conversation with someone else standing right in front of you.

It is the *absence* of that other person's voice that causes you to lean forward and possibly turn your head in the direction of the room where you last heard them speak. You were trying to determine if they were still there and if they were still speaking, but possibly at a lower level than you could hear. You were trying to *track a single voice* in the midst of all the other sounds and noise, listening for more of the same conversation from that same voice, and in the midst of all the other noise, you knew what to expect to hear. You were also trying to explain why you didn't hear them anymore.

6.2. Kinds of Knowing at the Party

The above is a variation on what is called "The Cocktail Party Problem" that has posed considerable problems for Artificial Intelligence. It contains many examples of things human beings normally know how to do that are found in Boundary Set *S*. Human beings have an ability to make objects of interest pop out from a cluttered, noisy background. We *know how* to isolate and excise objects from their background on the basis of many things, including the timing of the occurrences of things. We know how to form expectations about what we should look for and what to recognize next, without actually attending to or paying attention at all to what we are doing, or to even be aware that we are doing it. We ordinarily do not even *know that* we are doing this.

On lower levels of awareness that we do not ordinarily notice or even recognize, we make associations between both spatially and temporally contiguous information that is streaming on multiple levels, to our sensory and somatosensory-motor systems. Our brains know how to reinforce those connections between things that often appear together in its multiple, simultaneously occurring data streams that permit us to be aware of and recognize objects in our environment and in us. From all those streams of data on many levels, and in many complex interrelations, our brains form higher order combinations of synchronized features that permit us to make sense of our experience and act intelligently in the world.

The massive numbers of highly complex and interrelated neural networks in our brains process whole objects, such as a human face, knowing it is composed of eyes, nose, mouth and so forth, and that they nearly always appear together. Our brains are aware of and recognize higher-level objects, in part because of the similarities of their parts, to form an ascending hierarchy of related objects. The neural networks in our brains reinforce the interconnections among the parts of things, thus increasing our ability to detect and segment images into objects. The Cocktail Party Problem was intended to demonstrate that our real-world, real-time environment is very "noisy" and cluttered. Yet we know how to "track" single, unique objects in all that noise and clutter without much effort and without awareness that we are even doing so.

Moreover, as individuals, we experience the world around us and in us from our own points of view, with limited and often distorted perceptions of what is "out there" or "in us." We view a world in which there is a great deal that is often hidden from us by other objects, by lighting and darkness, and even by our own wishes and fears. Thus, our natural neural networks in our brains manage our intelligence by awareness and attending, on some level(s), to significant portions of occluded objects, thereby verifying or falsifying their presence in real-time.

6.3. Artificial Neural Networks

Let us look at some properties of artificial neural networks to assess whether or not they can handle the kinds of knowing found in the Cocktail Party Problem. First, for the sake of those readers who may be unfamiliar with artificial neural networks, I will review some very basic concepts, while keeping mathematical formulas at a minimum.

An artificial *neural network* is designed to model the way in which the brain performs a particular task or function. The human brain consists of countless cells called *neurons*. There are many different sorts of neurons, but they all share certain basic properties. The main cell body receives signals from other neurons by means of spindly extensions called *dendrites*. The neuron itself builds up a signal inside itself. When it reaches a certain threshold level, it "fires," discharging the signal down its long *axon* and over to other cells through connections at the end of the axon called *synapses*. These are connected to the dendrites of other cells. This model is simplified much more when it is simulated on a computer. Below is a graph of a computer neuron.

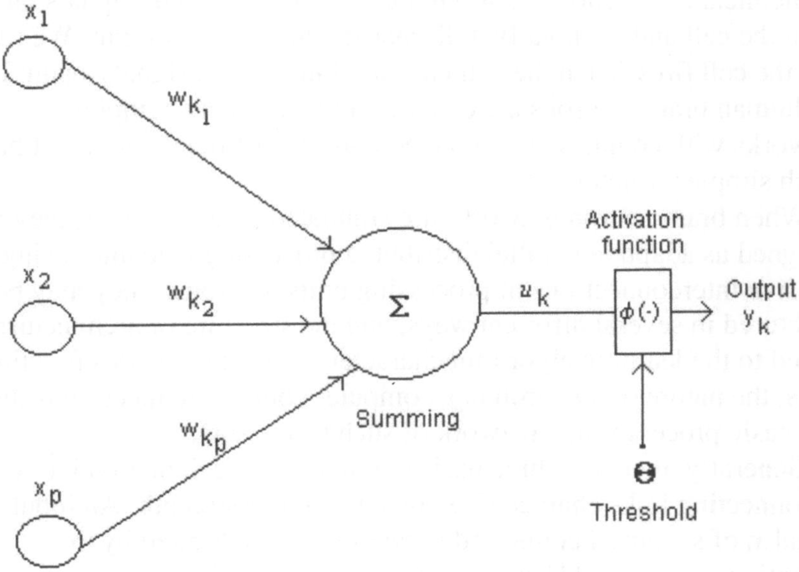

Figure SIX-1. Model of a Neuron

x_1, x_2, x_p = activation input signals
w_{k1}, w_{k2}, w_{kp} = synaptic weights of neuron k (vector of synaptic weights connecting the incoming neurons to the neuron of interest)
Σ = Additive function
Θ = Threshold activation function
u_k = linear combiner
$\varphi(\cdot)$ = activation function
y_k = output signal

The neuron is the circle at the center of the figure. It has several inputs which are the lines on the left (a real brain cell can have thousands of inputs) and a single output, the line on the right. The neurons in a network are connected together so that the outputs of some feed the inputs of others. The input to the cells arrive in the form of signals down the inputs. In the human brain, the signals take the form of chemical connections between the ends of nodes. The signals build up in the cell and eventually it discharges through the output. We say that the cell fires. Then the cell can start building up signals again. In the human brain, neurons are connected in extremely complex networks with countless interconnections. An artificial neural net has a much simpler structure.

When brain neuron networks are simulated on a computer, they are designed as adaptive parallel distributed processing machines using massive interconnections of processing units, neurons. They may be structured in several different ways, and the structure or architecture is linked to the learning algorithm characterizing the network's functions Thus, the nature of a neuron in a computer control architecture is that of a basic processor in a network of such processors.

Generally, there are three basic components. Each neuron has a set of connecting links characterized by a weight or strength. An input signal x_i of synapse j connected to neuron k is multiplied by the synaptic weight w_{kj}.. [The first subscript refers to the neuron in question and the second subscript refers to the input end of the synapse]. Each neuron also has a *summing function* for summing the input signals weighted by the respective synapses of the neuron. Each neuron also has an *activation function* [also referred to as a *squashing* function] which limits the amplitude of the output of the neuron. This function limits the permissible amplitude range of the output signal to some finite value, usually written as the closed unit interval [0,1] or as [-1,1]. An external *threshold function* is also included in the graph. This function has the effect of lowering the net input of the activation function. Thus, a neuron k can be described with the following equations.

$$u_k = \sum_{j=1}^{p} w_{kj} x_j$$

$$y = \varphi(u_k - \Theta_k)$$

Equations 1 and 2

In essence, each neuron sums its inputs with respect to weights, subtracts a threshold and applies an activation function to the result. The activation function, $\varphi(\cdot)$, defines the output of a neuron in terms of the activity level at its input. There are basically three types of activation function, including threshold and piecewise linear, but our concern in part is with *sigmoidal* ('S'-shaped) functions because they more closely characterize actual dynamic behavior of living things. Some form of nonlinear activation function is necessary in order to try to obtain dynamic, self-organizing mappings central to a mathematical characterization of kinds of knowing of interest here. It is especially important to realize that the self-organizing patterns of connections among the neurons of these networks is such that they are capable of modification *as a function of experience*. That is, neural networks can learn, and to some degree, they may be said to be able to come to know.

Haykin[1] has identified four different classes of network architecture: the single-layer feedforward networks; multilayer feedforward networks; recurrent networks; and lattice networks. I will briefly discuss each of these in turn.

6.3.1. Network Architectures

(a) A *single-layer* feedforward neural network is the simplest form of such networks. It consists of an input layer of source nodes or neurons that projects one-way [that is, it feedsforward] onto an output layer of computation nodes. The 'single-layer' refers solely to the output layer of neurons.

(b) *Multilayer* feedforward networks, which are the most common, contain one or more hidden layers of computation nodes called hidden neurons or hidden units. The purpose of the hidden units is to intervene between the external input and the network output. By adding one or more hidden layers, a network is enabled to extract higher-order statistics for a more global perspective in spite of its local connectivity with the extra set of synaptic connections and extra dimension of neural interactions. The outputs of neurons in each layer are connected to the inputs of the neurons in the layer above. The source nodes of the input layer supply elements of the activation pattern which constitute the input signals applied to the computation nodes of the second layer [that is, the first hidden layer]. The output signals of this second layer are then inputs to the third layer, and so on for the remainder of the network. The output signals of the output or final layer are then the overall response of the network to the activation pattern supplied by the source nodes of the initial input layer.

A network or ensemble is said to be *fully connected* if every node in each layer of the network is connected to every other node in the adjacent forward layer. If some of those communication links or synaptic connections are missing, then the network is *partially connected.*

Each connection from one node to another carries a strength which indicates how important the connection is. Strong connections have more influence on the node they connect into than weaker ones. They contribute more to the firing of the cell. The information carried by the network is stored in the differing strengths of the node connections. The strengths between the nodes are called *weights* in the program and are stored as numbers. Neural networks are being implemented in specially created integrated circuits, but most programmers simulate neural networks using software.

The term "feedforward" means that the connections between one layer and the next only run in one direction. There are connections from layer 1 to layer 2, from layer 2 to layer 3 etc. but no connections in the other direction. The opposite of a feedforward net is called a *recurrent* net, which have feedback connections.

(c) *Recurrent* networks have at least one feedback loop. They may consist of a single layer of neurons with each neuron sending its output signal back to the inputs of all the other neurons, with no self-feedback loops, or they may consist of a multilayer system of neurons with one or more feedback loops. Either single- or multilayer networks may also include self-feedback loops. Self-feedback is the output of a neuron fed back to its own input. Moreover, a recurrent network may have a hidden layer or it may not. As explained by Haykin, the presence of feedback loops in recurrent structures has a profound impact on the learning capability and performance of the network. Feedback loops also involve the use of particular branches composed of unit-delay elements which result in nonlinear dynamical behavior.

For our purposes, we should point out that recurrent networks often have attractor states, which we discussed earlier. This means that signals passing through the recurrent net are fed back and changed until they fall into a repeating pattern, which is then stable (*i.e.* it repeats itself indefinitely as it rattles round the loop). The input signals change until they reach one of these attractor states, and then they remain stable. When using recurrent networks, the goal is to train the weights so that the attractor states are the ones that you want.

(d) *Lattice structures* are one- or many-dimensional arrays of neurons with a corresponding set of source nodes supplying input signals to the array. The dimension of the lattice refers to the number of the dimensions of the space in which the graph lies.

6.4. Learning Algorithms

An artificial neural network is a massively parallel distributed processor that has a natural capacity for storing experiential learning and making that available for use. It learns or *comes to know* [in a restricted sense of 'know'] by means of algorithms which modify the synaptic weights of the network so as to achieve a learning or knowing goal. Generally, it is possible for neural networks to learn in a variety of ways and under certain specifiable conditions.

(a) *Supervised learning.* This occurs when there is an external "teacher", target or desired response the neural network is designed to achieve. The teacher has knowledge of the environment which is represented as a set of input-output examples to the network. The network, however, does not have knowledge of the environment.

Thus, the teacher or whatever makes up the teaching or desired response signal feed provides the network with the desired or target response, the optimum response to be performed by the neural network. Network parameters are then adjusted by both a training signal and error signal, until the network simulates the correct response. This in essence is referred to as the error-correction method of supervised learning. For our purposes, it is fundamentally flawed precisely because the environment is not in the feedback loop of the network. Moreover, without a "teacher" or teaching signal feed, a network cannot learn new strategies for particular or unique situations not covered by a set of defined examples used to train the network.

Other supervised learning algorithms include the Least Mean Square algorithm, which involves a single neuron [thus, it will not be of concern here], and the Back Propagation algorithm, which involves a multilayered interconnection of neurons. With this algorithm, error terms are back-propagated throughout a network layer by layer until it reaches the correct value.

(b) *Reinforcement learning*. This is learning of an input-output mapping by a process of trial and error for the purpose of maximizing a performance index, the reinforcement signal [see Haykin, 1994]. It may be either nonassociative or associative reinforcement learning. In the former, the reinforcement is the only input the network receives from its environment; in the latter, the environment provides information in addition to the reinforcement. With respect to associative reinforcement learning, with many kinds of input from the environment, it is necessary to carefully consider an *evaluation* function on the network, a *critic* function, and a *prediction* function.

A supervised learning system is one largely governed by a set of targets or desired responses. It is an instructive feedback system. On the other hand, a reinforcement learning system is one which is directed to improving performance and learning on the basis of any measure whose values can be supplied to the system. It is an evaluative feedback system. [2] In a supervised learning network, an external source [a "teacher"] provides direction to the system. In a reinforcement network, the network has to *probe*, that is *explore*, the environment through trial and error and delayed reward searching for directional information.

(c) *Unsupervised learning*. There is no external teacher or critic in this learning process. That is, there are no examples of the function to be learned by the network. Rather, a task-independent measure of the representation that the network must learn is used and the free parameters of the network are optimized relative to that measure. In effect, the network becomes "tuned" to the statistical regularities of the input data and develops the ability to form internal representations for encoding features of the input, creating new classes automatically.[3]

Self-organizing networks perform unsupervised learning. In general, these networks generate ordered mappings of the input data onto some low-dimensional topological structure, or they are used to partition input data into subsets or clusters such that the data inside one subset are similar [measured by the "nearest neighbor" principle or Euclidean distance, d^2], and items from different subsets are dissimilar. One way a neural network performs unsupervised learning is by means of a competitive learning rule in which neurons in a competitive layer of neurons compete with one another for the opportunity to respond to data provided in the input layer.

To some extent, artificial neural networks allow the characterization of a control system, for example the "brain" of a person or robot, as a set of layers of neurons and synapses. At least three layers are necessary because any network with fewer than three is limited in its ability to learn nonlinear mappings. Vectors and matrices are used to represent those layers. To attempt to generate the kind of self-organizing and dynamic knowing behavior of interest here, I will be concerned only with multilayer recurrent unsupervised networks with at least three layers. Though I cannot elaborate extensively upon the matter here, robotic control systems and those mathematical configurations appropriate to them are obviously germane in the characterization of *knowing how* because of the bodily capacities involved.

6.5. Multilayered Synchronous Networks and Self-Organization of Boundary Set *S*

With respect to possible computer generated characterizations of Boundary Set *S*, the issue actually comes down to mapping some variety of unsupervised multilayer recurrent neural network topology onto random Boolean networks. Since the Boundary Set *S* is a complex self-organizing dynamic transactional set, we are not addressing any supervised learning model such as backpropagation, or BAM (bidirectional associative memory), or Boltzmann models. This is so because supervised learning algorithms for multilayer neural networks, for example, have at least two serious problems: they necessitate a teacher [teaching signal] to specify the desired output of the neural net, and they necessitate a method of communicating error information to all of the connections.

Of course, where there is no external teaching signal to be matched, it is necessary to specify some means to force the hidden units to extract the underlying structure in the raw inputs. Moreover, set *S* is a *transactional* epistemic set because of the nonlinear dynamic and *emergent* properties one finds in the global properties arising from that set's local interactions (in part) with whatever constitutes its environment. Thus, we need feedback loops providing information to the network from the environment, making it a recurrent network architecture.

The primary practical advantage of artificial neural nets is the very directness of the model. For example, with respect to generating *recognizing* [in its representational sense] behavior, say the recognition of highly differentiated forms of the same alphabet character, we simply map pixilated images to alphabetic characters. The pixels of an image are neurons and are connected to output neurons. The usual GOFAI approach to the problem would be an indirect statistical model of learning correlations. That is, GOFAI would require an extensive up front program with a database of all known factors needed to make correlations between images and character. This approach, however, came to a standstill in fields of pattern recognition, clearly central to generating characterizations of behavior found in Boundary Set *S*, voice and handwriting recognition, and others.

Artificial neural nets present a superior method of taking data presented and determining what data is relevant to a task, and improving the performance of that task in some way. In the above recognition example, I noted that the pixels of the image are neurons and are connected to output neurons. The output neurons are the ones that identify the alphabet character. However, the connection between the image neurons and the output neurons is likely to be through at least one hidden layer of neurons. It is this layer which is in fact a *multilayered level of neurons*, a hierarchy of primitive epistemic elements which are categories [relations] within immediate awareness which we may try to simulate. It is at this hidden layer of a network of neurons where particulars making up data from all parts of the input layer must be combined at individual neurons, according to Boolean functions.

6.6. Self-Organizing Neural Networks

A self-organizing system is an unsupervised one. The goal of a learning algorithm in a self-organizing neural network is to discover and reproduce significant patterns or features in the input data [particulars], without a teacher or external teaching signal feed. In essence, the algorithm has a set of local rules by which changes to synaptic weights are confined to neurons in the immediate neighborhood of those whose weights are changed. This allows it to learn to compute input-output mappings with certain properties. Generally, learning takes place according to some Hebbian or competitive rule, by which there is a repeated modification of the weights of all the connections in the network in response to activation patterns until a final configuration develops. Though self-organization can be found in feedforward networks with single layers, for our purposes, I will focus solely upon a feedforward network with multiple layers in which the self-organization proceeds on a layer by layer basis. Moreover, I will restrict the discussion to competitive learning, specifically that class of maps called the self-organized feature map (SOFM) or vector coding algorithms.

As noted earlier, vectors and matrices are used to represent layers of a neural network, and for nonlinear mappings we require a minimum of three layers. Thus, with respect to generating Boundary Set *S*, using the above primitive categories of the relation of immediate awareness [within sign relations], we might generally try to conceptualize the network as suggested in the following multilayered topology which has an unsupervised learning capability, with a sizable hidden layer:

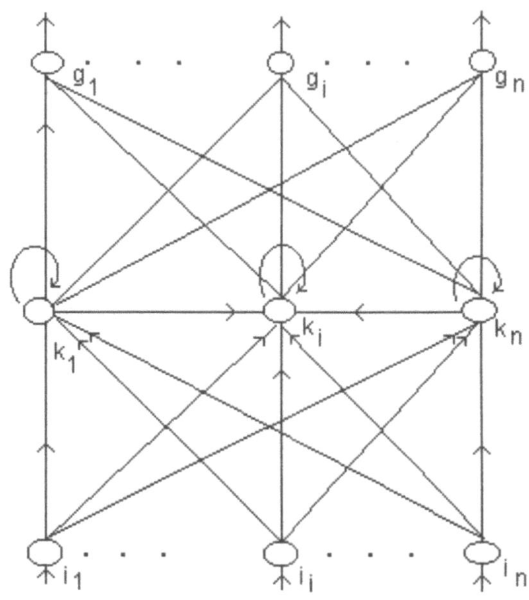

Figure SIX-2. Counterpropagation Model

The above graph shows a representation of a feedforward network with lateral feedback and with one layer of self-feedback loops. It is fully connected in the sense that every node in each layer of the network is connected to every other node in the adjacent forward layer.

The function of the hidden neurons is to intervene between the external input and the network output, and it is precisely where we should expect to find the set of primitive relations of immediate awareness. With one or more hidden layers, the network acquires a *global* simulation or perspective in spite of its local connectivity. The global simulation is necessary to generate the macroscopic levels of knowing performance sets, of the kind we find in ordinary knowing how to do a thing.

Moreover, the network architecture must be intended to parallel process a virtually *uncountable continuous domain* of input particulars presented to it. These are the sensory and somatosensory-motor particulars which are the *terms* of the primitive immediate awareness relations. In contrast to classical symbolic processing of AI, the intention here is to present an intelligence architecture which is *massively* parallel in processing of information. The computation of the network is spread over many neurons and it will not matter much in terms of the actual knowing behavior generated whether or not the states of some neurons are not in accord with their expected values. Noisy particulars, such as we have discussed above where the object [particular] is not entirely distinguished from its ground, may still be recognized by the network. Moreover, even damaged networks will still function, though perhaps not perfectly. The performance of the network will degrade gracefully within a certain range.[4]

As Blum[5] notes, counterpropagation models such as the above are actually a combination of two other models, the Kohonen linear vector quantizer and the Grossberg outstar encoder, though the capabilities of the counterpropagation model extend beyond the capabilities of these two. The Kohonen layer (indicated in the above model with 'k') demonstrates the generalization and "lookup table" capabilities of self-organizing networks. The Grossberg layer (indicated in the above model with 'g') demonstrates the outstar's ability to act as a minimal pattern encoder. As noted, counterpropagation models have a capability to handle data [particulars] that have not been seen [presented] before, as well as data which are erroneous, and they exhibit the property of self-organization. But the models handle the data not seen before by means of a principle of similarity, and hence categorize or classify input patterns by means of that principle.

As the above model shows, counterpropagation models are three-layered networks, consisting of the input pattern, hidden layer, and output pattern, with two layers of synapses. This model presents an unsupervised learning model because the weight adjustments on synapses are not made based on any comparison with a "target" output. With each input pattern, the hidden layer neuron whose vector of synapses to the input layer is most similar to the input pattern is chosen as the selected ("winning") neuron. The second layer of synapses, indicated in the above model with 'g', has weights adjusted on each pattern association based upon the activation of the winning hidden layer neuron and output vector. In sum, the counterpropagation model is a competitive learning model.

But a counterpropagation network model cannot process unique particulars *as unique*, because processing the input pattern is based upon a principle of similarity[6] of the vector of synapses of the hidden layer to the input pattern. The greater the similarity of the input pattern to the vector of the synapses of the "winning" neuron, the greater the weight adjusted on each pattern association, thus categorizing [classifying] the input pattern. With respect to *natural intelligence* systems, this brings us back to issues raised earlier: (1) That primitive relations of immediate awareness are not class objects; and (2) the fact that kinds of knowing found in Boundary Set *S* are not reducible to *knowledge that*, our knowledge of classes.

Nonetheless, such a model is very well suited to discover statistically salient features useful for classifying a set of input patterns. But it is not sufficient [though it may be a necessary component] in a mathematical model for immediate awareness, which depends upon combinations of many, perhaps an immense number of primitive epistemic relations at any given moment in time.

6.6.1. Self-Organized Feature Map (SOFM)[7]

The last artificial neural network model I will look at is Kohonen's Self-Organizing Feature Map (SOFM), though a thorough exploration of SOFM is beyond my purposes here. I want to generally review certain of the properties of the model and look at some of its limitations as well as possible uses. As explained by Haykin, in a self-organized feature map, neurons usually are placed at the nodes in a one- or two-dimensional lattice, though higher dimensions are possible. As earlier noted, these neurons over time become selectively "tuned" to input patterns, that is vectors, in the competitive learning process. Those "tuned" neurons are the "winning" neurons referenced above.

SOFM is unlike most of the other architectures presented earlier because it involves *unsupervised* training. It learns patterns for itself. In *supervised learning*, networks are trained by presenting them with an input pattern together with a corresponding output pattern, and the networks adjust their connections so as to learn the relationship between the two. However, with the SOFM, we just give it a series of input patterns, and it learns for itself how to group these together so that similar patterns produce similar outputs.

As Haykin explains, a self-organizing feature map is characterized by the formation of a topographic map of the input patterns in which the spatial locations, that is coordinates of the neurons in a lattice, correspond to intrinsic features of the input patterns. It is precisely this characteristic of SOFM which makes it useful as a mathematical characterization of certain of the primitive epistemic relations of *immediate awareness* embedded within *knowing how*. The topographic computational mappings of SOFM follow the topographically ordered computational mappings and organizational structure of the brain:

> In particular, sensory inputs such as tactile. . .visual. . .and acoustic. . .are mapped onto different areas of the cerebral cortex in a topologically ordered manner. . .the neurons transform input signals into a *place-coded probability distribution* that represents the computed values of parameters by sites of maximum relative activity within the map.[8]

The computational maps of the brain are the different sensory, motor, and somatosensory parallel processing neuronal networks mapped onto corresponding areas of the cerebral cortex. They are subject of course to representations showing the frequency distribution of neuronal firings of the sensory and somatosensory systems in an active knowing system.

We can to some degree use those maps for our epistemic primitive categories, assuming that those primitive relations are also topologically mapped correctly.[9] Thus the *place* of the information-bearing signals, that is the epistemic terms in a primitive epistemic relation [which may be dual-, triple-, quadruple-, and so on] is a term-place within a relation complex. For example, the primitive relation of *tasting* might be the following many-term relation of relations which are themselves terms of a complex: Tasting (attending (feeling [tactile] (smelling (s,o)))). Such a many-term, multilevel set of primitive relations will to some degree amount to an array of *synchronous* neural activity across the appropriate parts of the multilayer sensory and somatosensory networks. The entire array of the topological ordering of mapped information signals [primitive terms of relations] can then be accessed by higher-order processors using simple connection schemes, thus resulting in global effects. Global behavior thus emerges from many local interactions.

Computational maps in the brain and in general are central to understanding a self-organizing multilayer neural network architecture designed for the purpose of self-building of topographic maps. This would in turn be central to any computer architecture of Boundary Set *S*. As with the nervous system, computational maps can continuously analyze very complex events occurring in a continuous dynamic environment which requires parallel distributed processing strategies for handling virtually uncountable amounts of information. Moreover, as Haykin notes, these maps simplify the schemes of connectivity required to utilize the information by higher order processors. There is a common mapped representation of the results of different kinds of computations. This permits a *single* strategy for making sense of the information. Lastly, representations of features in topographic form enable the fine-tuning of a basic processor not otherwise possible.

Kohonen's map basically attempts to capture the *essential* features of computational maps *in the brain* and is more general than previously presented self-organizing maps because of its data compression capabilities [the reduction of dimensionality on the input]. As Haykin notes, it belongs to a class of vector coding algorithms because it provides a topological mapping that optimally places a fixed number of vectors into a higher dimensional input space thereby facilitating data compression. With respect to a possible SOFM characterization of Boundary Set *S*, I will make reference to the sensory relations of the of primitive immediate awareness relations found in the schema presented in the last chapter.

Fundamentally, the aim of the SOFM algorithm is to store a large amount of input vectors by finding a smaller set of prototypes in order to provide a good approximation of the original input space, *X*.[10] Given an input vector *x*, the SOFM algorithm first identifies a best matching or winning neuron *i*(*x*) in the output space A, in accordance with the feature map, **Φ**. Where *X* denotes a spatially continuous input (sensory and somatosensory) space, A denotes a spatially discrete output space, and **Φ** denotes a feature map [a nonlinear transformation map of input space *X* onto output space A], then

$$i(x) = \arg_j \min \|x - wj\|, \; j = 1, 2, \ldots, N.$$

Equation 3

We can use the index *i*(*x*) to identify the neuron that best matches the input vector *x*, and can determine *i*(*x*) by applying this equation. It is the value of *i* that we want, and the neuron that satisfies this condition is the *best matching or winning* neuron. The equation describes the best matching criterion which is equivalent to the minimum Euclidean distance between vectors.

By using the above equation, a continuous input space is mapped onto a discrete set of neurons. Depending upon the application, the response of the network could be either the index of the winning neuron [position in the lattice] or the synaptic weight vector closest to the input vector in a Euclidean sense. For example, the input space *X* may represent a coordinate set of somatosensory receptors distributed over the entire body surface. The output space could represent the set of neurons located in that layer of the cerebral cortex to which the somatosensory-motor receptors are confined. With these mappings, and with an expanded set of input categories defining the input space to include our primitive epistemic relations of Boundary Set *S*, we can in principle obtain mappings of kinds of *knowing*.

Figure SIX-3. SOFM: Relationship Between Feature Map and Weight Vector

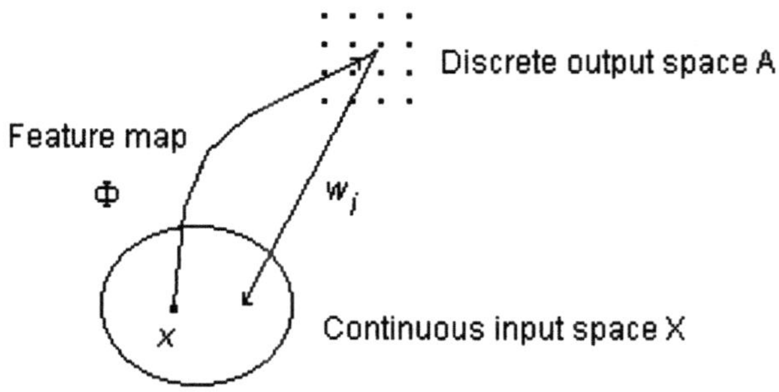

The SOFM is generally speaking a two-dimensional structure, a rectangular grid of cells, which allows activity to appear on its surface and represent groupings. The feature map, **Φ**, displayed in the input space *X*, is self-organizing and represented by the synaptic weight vectors $\{wj \mid j = 1, 2, \ldots N\}$, and can provide a good approximation to the input space *X* in the output space A. The feature map approximates the input space *X* by pointers or prototypes in the form of synaptic weight vectors, w*j,* such that the feature map provides a faithful representation of the important features characterizing the input vectors $x \in X$ [Haykin, 1994].

A major property of SOFM is the topological ordering of the feature map **Φ** such that the spatial location, the coordinate, of a neuron in the lattice corresponds to a particular domain or feature of the input patterns from input space *X*. This is accomplished by an update equation which forces the synaptic weight vector w_i of the winning neuron *i*(*x*) to move toward the input vector *x*, and to move the synaptic weight vectors w*j* of the closest neurons *j* along with the winning neuron *i*(*x*).

Thus feature map **Φ** is a net with nodes having weights as coordinates in the input space *X*. If we can simulate it at all, the feature map **Φ** of Boundary Set *S* would consist of nodes identified with each of the primitive epistemic relations cited in the above list [without abstract primitive objects], including the sensory and somatosensory relations, as well as the more problematic relations of moving and touching.

6.7. Adaptivity

The SOFM is self-organizing precisely because the synaptic weight vector w_j of neuron j changes in relation to input vector x. It does so in accordance with a modified Hebbian rule whereby a nonlinear forgetting term is introduced to overcome the problem of saturation of weights. The latter problem is due to changes in connectivities under the basic Hebbian rule occurring in only one direction.

In essence, Kohonen's SOFM is a simple geometric computation. After initialization, there are three basic steps in its application: *sampling, similarity matching, and updating*. The sample x should be drawn from the input distribution with a certain probability, and vector x is a sensory signal. Similarity matching is done as I outlined above, that is by finding the best-matching or winning neuron $i(x)$ at time n, using the minimum-distance Euclidean criterion stated above. Then the synaptic weight vectors of all neurons are adjusted by using the update formula also cited above. The final step is simply to continue with the second step until no noticeable changes in the feature map are observed. Over time, the feature map will reflect variations in the statistics of the input distribution. That is, vectors x with a higher probability of occurrence will be mapped onto larger domains of output space A and with better resolution than those vectors with low probability of occurrence.

There are other facets of SOFM which could be discussed in a more detailed examination of the algorithm and its potential use for simulating or generating elements of Boundary Set S, for example the addition of a conscience to the competitive learning mechanism. In the interests of brevity, I will leave such discussions for others.

6.8. Critique of Artificial Neural Network Models

The only unsupervised, self-organizing model we considered was Kohonen's SOFM. All the others have one or another disadvantage, mostly due to the fact that they are supervised. An advantage of Kohonen's model is that it is motivated by neurobiological considerations, specifically by the computational maps designed to perform in general the way the brain performs. But it does not handle hierarchical structures very well, in part due to problems with classifying data and forming categories. It is not successful in extracting or organizing hierarchical categories from streams of sensory data, such as the categories of primitive relations found in immediate awareness. Recall the cocktail party problem and the kinds of knowing found there. Such streams are often highly cluttered and noisy, containing a lot of irrelevant information that must be "selected out" in order to simulate kinds of knowing found at the cocktail party. Indeed, it appears that much *prior* categorization is necessary to get the SOFM to operate well, and it certainly does not simulate how humans are able to "track" single voices and be aware of the *absence* of single voices in a noisy crowd. The necessity for prior categorization, which would not be possible to computationally solve the cocktail problem anyway, is clearly a major deficit with respect to an architecture necessary to simulate kinds of knowing found in Boundary Set *S*.

Moreover, any adequate model to map such real time streams of sensory data must have an architecture that memorizes the synchronicity among those various sensory inputs. It must score each new sensory experience for its similarity to all previous such experiences, which the SOFM may be able to do to some extent, but it must form higher order categories based on those.

Such a model must also automatically recognize higher-level, emergent objects by logging similarities of component parts to form a hierarchy of related objects. It must have a working memory that categorizes associations by similarity of objects. With a face, for instance, it must be able to self-organize a higher-level face object with the component parts. By reinforcing associations between spatially and temporally contiguous information, it must be able to reinforce those connections between things that appear together and often in its data stream. Due to those reinforced connections, it must be able to form higher-order combinations of synchronously occurring features. These must occur from the topology of the network. But SOFM is not adequate to do much of any of these things. For that very reason, even if other problems did not exist, it cannot simulate Boundary Set *S*.

But what about indexicals? Can *any* computer handle those? One of the ways AI engineers and others try to enable a computer to behave as speakers of a natural language behave is to design a meaning representation language for that natural language. The natural language is then translated into the meaning representation language which in turn is then used by the computer to generate natural language behavior.

6.9. Natural Language Semantics and Indexical Reference: More Limits of Computation

A widely used method to translate natural languages into a meaning representation language is Zadeh's Test-Score Semantics.[11] It has been used as a method of translating natural language propositions into PRUF, a meaning representation language based on fuzzy logic.

To its credit, Zadeh's Test-Score Semantics (TSS)[12] and PRUF are preferable to classical theories of meaning representation of natural language precisely because of the non-standard assumptions on which that theory is built. Zadeh rightly assumes that no mathematical theory based on two-valued logic is capable of characterizing the elasticity, ambiguity, and context-dependence which set natural languages far apart from the models associated with formal syntax and set-theoretic semantics. This is due to the fact that issues of grammaticality and certainly issues of meaning in natural languages, are almost all a matter of degree.

Hence, as an alternative to approaches based on two-valued logic, PRUF provides the concept of a *possibility distribution* as a natural mechanism for the representation of much of the imprecision and lack of specificity intrinsic to natural language communication among human beings. Possibility theory is distinct from the bivalent theories of possibility related to modal logic and possible world semantics. And *possibility* distribution is analogous to, yet also distinct from, that of a *probability* distribution. But among other problems, PRUF shares the same fundamental flaw that all meaning representation systems share: it conflates grammatical meaning with mathematical function.

To some extent, PRUF makes possible the representation of the meaning of propositions containing fuzzy quantifiers such as 'many', 'most', 'few', 'almost all'; and modifiers such as 'very', 'more or less', 'extremely'; also fuzzy qualifiers such as 'quite true', 'not very likely', and 'almost impossible'.[13] The semantics theory underlying PRUF is test-score semantics in which the concept of an aggregation of test scores is central. It is broader in scope and hence subsumes most of the semantical systems which have been proposed for natural languages, including both truth-conditional and possible-world semantics.

In Test-Score Semantics (TSS), an entity in natural language discourse such as a predicate, proposition, question or a command generally has the effect of inducing elastic constraints on a set of objects or relations in a universe of discourse. As Zadeh explains, the meaning of such an entity may be defined by: (i) identifying the constraints which are induced by the entity; (ii) describing the tests that must be performed to ascertain the degree to which each constraint is satisfied; and (iii) specifying the manner in which the degrees in question or the partial test scores are to be aggregated to yield an overall test score. In essence, the meaning of a linguistic entity in a natural language is identified with the implicit or explicit constraints on the entity.[14]

We can assess how well TSS works for translating certain indexical entities of natural language discourse into PRUF, closely adhering to Zadeh's rules and descriptions of the use of the procedures for Test-Score Semantics in our assessment. If the general goal of PRUF is to eventually achieve machine translation of a wide variety of expressions in a natural language, then we must look to *ordinary everyday discourse* for examples of language for dialogue. Close analysis of the semantics of such discourse will provide a clearer picture of the current limitations and necessity for extensions of PRUF. This is especially the case since PRUF is used as a basis for question-answering systems in which the knowledge base contains imprecise data of the sort referenced above, predicates such as 'young', 'big', and quantifiers such as 'many', 'most'.

But natural language used in everyday discourse of a question-answering type also almost always includes indexical expressions such as 'this' and 'that'. For example, while pointing to a far distant white dot on a landscape, one person A says to another, B, "That is Richard's house." The apparent simplicity of such statements and the use of the demonstrative indicators or indexicals 'this' and 'that' can be very misleading and present apparently intractable problems for a meaning representation language such as PRUF and Zadeh's Test Score Semantics (TSS).

In examining those problems there are a number of very weighty issues to be touched upon, and not all may be pursued as thoroughly as they might be. I will consider both thinking and speaking, as well as speaker's and hearer's meaning, and how semantic rules determine the content of both, with emphasis on indexical reference. For the time being and because I have addressed these earlier, I will also set aside more intriguing questions such as those regarding whether in thinking the mind can establish a direct connection with, or immediate awareness of, what it thinks, or whether the mind must always think its objects by means of the mediation of linguistic concrete or abstract representations.

It has been a fundamental thesis throughout earlier portions of this work that I hold that symbolic, linguistic representations are not necessary for the mind and that the mind does have immediate primitive relations with objects of awareness. For purposes here, however, I will assume thinking is symbolic. I will assume that thinking is a process somehow conducted solely by means of symbolic tokens or representations in the mind. I will also assume that what a natural language speaker says and thinks is at least partially or fully determined by the semantics of the tokened symbols that embody or constitute his or her speaking and thinking. These assumptions will be for the express purpose of determining whether or not this view of thinking actually holds together. If we find any instance where it cannot account for our *ordinary* everyday language and thinking, then my arguments pose a serious challenge to the strong AI thesis that we are all digital computers, that all thought is computational.

I am primarily concerned here with the indexical meaning and use of indicators such as 'this' and 'that' and will focus on their meanings by way of their pragmatic properties. The *pragmatic* properties of indicators are those they have by virtue of being used in accordance with general rules or purposes independently of whatever denotations or syntactical properties they may have or acquire in contexts of use. *Syntactical* properties of indicators are word order and logical scope in the sentences containing them. *Semantical* properties are those indicators have by virtue of how they pick out denotations in particular occasions of their use.[15] In general, it is useful to think of indexical reference as a *procedure* through which a language speaker *picks out or points to* aspects or pieces of reality for perspicuous attention or confrontation. A token of a public (visible) demonstrative refers to a material object, for example a person, rock, shadow. To test Zadeh's TSS and PRUF, consider the following ordinary, normal visual experience described in the following situation:

> Richard K. has invited some friends for weekend discussions at his summer residence in the Davis Mountains in West Texas. Mary L. is driving some of us. Suddenly, just as we round a bend on Apache Mountain, Mary L. slows down, and says (not pointing):
>
> (1) "That tiny yellow dot on the right is Richard's house, where he stays in the summer." We all turn our heads to the right, looking for a tiny yellow dot. With a smile, Mary L. quickly points out that statement (1) does *not* entail that:
>
> (2) "Richard lives in a tiny yellow dot."
>
> In support of Mary L.'s claim that there is no implication, Paul B. explains:
>
> (3) "Richard's house, where he stays in the summer, is the same cottage built by Count von Erstberg in 1803."[16]

The above (3) shows us that the properties of being a tiny yellow dot, that is being a dot, yellow, and tiny, do not belong to Richard's house. That is, they are not properties of a physical object. Mary L.'s *that* in (1) refers to an item *in her visual field* and it is *that* item which has these properties.

Statement (1), however, has the following form:

(1A.) (a)[That which is a tiny, yellow dot is] the SAME as (b)[Richard's house where he lives in the summer].

But the sameness asserted here is *non-commutative* in the sense that not all properties transfer across, from (a) to (b).

The fact that (1) above does not entail (2) is confirmed by (3). The non-commutative sameness of (1), however, contrasts with the sameness of (3) which is *commutative:*

(3A) (a') [Richard's house where he lives in the summer is] the SAME as (b') [the cottage built by Count von Erstberg in1803].

In (3A), properties transfer across (a') and (b') and the terms that are the same inherit SAMEness partners. That is, (1) or(2) and (3) above imply:

(4) That which is a tiny, yellow dot is the SAME as the cottage built by Count von Erstberg in 1803.

Some of the obvious conclusions to be drawn from the above discussion are the following:

(a) Ordinary, objective, *intracategorial* true identities are commutative in the sense that all properties transfer across the sameness, as for example as in (3A);
(b) Mary L.'s *that* (a tiny yellow dot in her visual field) is *not* identical with Richard's summer house;
(c) Hence, Mary L.'s visual demonstrative *that* does not immediately pick out the physical entity, Richard's house;
(d) Mary L.'s tiny yellow dot is not a real dot [but a subjective item in her visual field].

But a startling conclusion which we must also draw is the following:

(e) From the above, it appears that we build perceptual judgements, in part, *on discounted linguistic and conceptual illusions.* These discounted illusions consist in the following:

(i) That the word 'is' in (1) and (3) expresses the same *sameness;*
(ii) That Mary L. uses 'that' to refer only to a physical object, Richard's house;

Mary L. refers to Richard's house with *'that'* only in a derivative, broad sense in which she claims that her visual *that* is the same as Richard's house. Hence, we also conclude:

(f) Sameness is a *mixed* relation, equating a *subjective* item (in a visual field) with something *objective,* a physical object. It is transcategorial, hence non-commutative.

This example is a very rich source of problems for meaning representation languages. It reveals layers of meaning and thought, including discounted illusions, that often accompany our everyday, ordinary natural language use of indexicals. But let us see what Zadeh's TSS and PRUF can do with it.

Utilizing TSS, and the above example on the indexical use of *that,* we want to assess the effect the indexical word 'that' has on inducing elastic constraints on a set of objects or relations in a universe of discourse. Hence the meaning of the word 'that' will be defined by identifying those constraints, describing tests that must be performed to ascertain the degree to which each constraint is satisfied, and specifying the manner in which the partial test scores are to be aggregated to yield an overall test score.

As Zadeh explains,[17] the conceptual framework of TSS is rooted in our intuitive perception of meaning as a collection of criteria for relating a linguistic entity to its object, its designation. That is, there must be some external object (World) or proposition that is the designation or "target" of a linguistic entity. Using the above description of a complex indexical situation, we want to test whether or not Paul B. understands the meaning of proposition *p*, (1), uttered by Mary L.: *p* = "That tiny starchy white dot on the right is Richard's house, where he lives in the summer."

Following Zadeh's TSS, we might present Paul B. with a variety of scenes or worlds (state descriptions or databases) depicting or describing the joint action of Mary L. relative to Richard's house, then asking Paul B. to indicate for each scene or world W the degree, c(W), to which W corresponds or is compatible with Paul B.'s perception of the meaning of *p*. Additionally, if he can articulate the tests which he performs on W to arrive at c(W), then we would conclude that Paul B. not only understands what *p* means ostensively, but he "can also precisiate the meaning of *p* by a concretization of the test procedure."[18]

In contrast to two-valued truth-conditional and possible world semantics, TSS permits the degree of correspondence or compatibility, c(W), to be any point in a linear or partially ordered set, *e.g.* the unit interval [0,1]. Replacing scenes or worlds with databases, assuming Paul B. is presented with the proposition *p* and with a database D, and he performs a test, T, on D, this yields test score $t : t = T(D)$; $t =$ Comp. (*p*, D), where T is a representation of the meaning of *p*, and *t* is a measure of the compatibility of *p* and D.[19] Of course T is composed of a set of tests, $T_1, T_2 \ldots T_n$, where the overall test score *t* is the aggregation of constituent test scores $t_1, t_2, \ldots t_n$. Moreover, TSS permits either a single test score, that is a number in the interval [0,1], or *t* may be a vector, $t = (t_\alpha, \ldots t_\gamma)$ where each component is a number in the interval [0,1] or a probability/possibility distribution over the unit interval. We can represent the world (or database) as a collection of relations, *i.e.* a relational frame which includes the name of the relation and names of variables are included as columns as well as table entries.

Below is the relational frame of the relation 'sameness' as applied to the above example:

SAMENESS	COMMUTATIVE	NON-COMMUTATIVE
1A	Empty	√
3A	√	Empty

1A: *That* which is a tiny, yellow dot is the SAME as Richard K.'s house where he lives in the summer.

3A: Richard K.'s house where he lives in the summer is the SAME as *that* huge, blue, 4-story villa built by Count von Erstberg in 1803.
√ : obtains

In terms of the categories of commutative/non-commutative, the relation of sameness is mutually exclusive: it is either commutative with respect to 1A and 3A or it is not. However, this relation does not reflect the natural language use of 'that' in this context. As noted above, Mary L.'s proposition (1) and Paul B.'s proposition (3) are built on linguistic and conceptual illusions that are part of the natural language dialogue.

Nonetheless, the dialogue is entirely coherent and each sentence (and each word in each sentence in the order in which they appear) functions appropriately, perhaps because of the linguistic and conceptual illusions, which (perhaps on some meta-transformational semantical level) are discounted by both speaker and hearer. TSS could not be used to test illusions that are nonetheless part of an ordinary everyday use of natural language, hence it could not be used to translate these illusions into any meaning representation language. TSS cannot handle ordinary, everyday language indexicals.

Furthermore, Mary L. makes clear that (1) does not entail (2), that the tiny yellow dot on the right which is Richard's house does not entail that Richard lives in a tiny yelllow dot. Not only must an adequate meaning representation language provide translation rules sufficient to characterize discounted linguistic and conceptual illusions on which natural language dialogue is sometimes based, it must also provide for *non-implications* of the above kind.

6.10. The Conflation of Grammatical and Indexical Meaning with Mathematical Functions

Certain of the problems faced by Zadeh's TSS center on the conflation of grammatical (syntactical-semantical) meaning, especially indexical meaning, with mathematical functions, sets of ordered *n*-tuples. But grammatical meanings are kinds of structure present in contents of *thought*, hence they cannot *literally* be mathematical functions. There appear to be two major assumptions underlying such formal-semantic modeling of natural languages:

(1) In a context C of natural language about a domain U of entities there is always a function *i* mapping the set T of singular terms in a piece of natural language used in C to the entities in U: function *i* : T, C $\rightarrow$ U;

(2) The function *i* posited is constituted by the grammatical meanings of T, the syntactical and semantical meanings of T, (and in the case of sentences asserting propositions, the function is constituted of fuzzy values or truth-values).[20]

Each grammatical meaning *t* in T : C $\rightarrow$ U. That is, the syntactical-semantical meaning of an expression, including the cognitive content of an expression, is a function that maps contexts of speech into speech contents. Functions such as that cited under (1) are apparently operative in veridical thinking. Such is the case in the above perceptual judgment in (3) and (3A). As Castañeda points out,[21] it may be perfectly appropriate for formal semantics or those concerned with issues of veridicality to ignore the psychological mechanisms [I would say primitive immediate awareness mechanisms or structures] that actualize the mappings. But those concerned with the functioning of natural language cannot ignore those mechanisms.

However, the function *i* in (1) in a natural language situation or context must be understood in the following way: In a natural language situation or context C, there are factors which cause the thinker/speaker to think the (one) appropriate functional value, *not the function itself.* Interpretive functions *i* are *external* to the cognitive content of thinking,[22] though they are obviously needed in theories *about* the processes of thinking and veridicality.

But as we saw above in (1) and (1A), a basic interpretive function *may fail to operate* in illusory linguistic and conceptual experiences. In (1) and (1A) there is no proposition that is the external target of Mary L.'s statement "That tiny white dot is Richard's house." Mary L. speaks (and thinks) demonstratively. And even though a functional argument is available, the functional value is missing. Zadeh's thesis (2) hence, must be rejected. That function *cannot* be cognitive content and neither can it be the same as grammatical meaning.

It may be instructive to explore a bit further what a function is supposed to be as it pertains to a set-theoretical mapping of natural language grammatical meaning to a domain U, for example, a mapping into the unit interval. Mathematical functions can be sorted into several categories. For some finite function, we can think of an ordered pair in *propria persona,* such as counting the number of pencils on my desk. Other finite or infinite functions we cannot think in *propria persona* but as falling under a property or set of properties which we can then think in *propria persona,* for example $f: A \rightarrow B$ where A is the set of natural numbers N and B is the set of all integers Z. We have a complex property representing the infinite function f, that is we have a rule for finding pairs of arguments and values of the function. But we are not thinking the function in *propria persona.* As pointed out by Castañeda,[23] in theory of language we must distinguish between the following two kinds of proposition: (a) The quadruple of 3 is a number; (b) $F: N \rightarrow \{y: y = 4 \times x \text{ and } x \in N\}$ and $F(3)$ is in N. Proposition (a) is a simple relational proposition; proposition (b) is a functional proposition. The two are equivalent but not identical. To think (a) is not to think (b).

If cognitive content (or fuzzy and truth values) and grammatical meaning are mathematical functions, then they are sets of ordered pairs. But this is not what Mary L. thinks or says when he thinks and says "This is Richard's house," nor "This appears to be Richard's house." In ordinary natural language discourse, there is no conscious thought of the items which constitute the speech context (as arguments), and then an assignment of a value to that argument. That is, there is no conscious inference from the function and context to *this.*

But there is an even more serious problem with any approach which conflates indexicals with mathematical functions. Looked at as a system, we could count the number of parts or variables which comprise the function which could conceivably be the character of 'this'. The algorithm which would result would be practically unwieldy in size. For example, if we assume the function which is the character of 'this' to be the following,[24] we can demonstrate the unwieldy results even from a non-statistical point of view:

f: {(a, B, t, p, s (this), D) : where a is a speaker, B the domain of hearers, t the time of utterance of an English sentence, s ('this'), with the demonstrative 'this', p the place at which a speaks, D a domain of objects referred to by *a* in his utterance of *s* ('this') lying in the vicinity of *a* }$\rightarrow$ D.

To determine a set of possible effects, an equation for each part in isolation, one for each combination of parts, and one for the context are required. For example, for a system of two parts only four equations are required, but for one of ten parts the number of equations increases to 1,035. In the above natural language discourse context, assuming the function captures 'this' as a proximal demonstrative of English (which is in doubt, as I will mention below), we have a system of 6 parts. Hence we need an equation for each part in isolation (6), one for each combination of parts, that is for *n* parts, we need 2^n combinations or 2^6 combinations, (or 64 equations), for a final total of 71 equations.

Obviously, the growth in the number of equations comes from the possible combinations of parts. For a system of 20 parts, there would be 2^{20} or over a million combinations. Hence a function with 20 parts or variables is simply too large for non-statistical treatment. Alternatively, we could view the function from a statistical point of view. That is, in place of accounting for each combination or interaction of parts or variables of the function, we could treat average combinations or interactions. Hence we can shift from absolute values to probable ones. But this presents problems as well. To secure accuracy the system must be very large. The relative error of average values is of the order $1/\sqrt{n}$. Above, a function with 20 parts is too large for non-statistical treatment, but it is too small for the statistical approach since it would permit an accuracy of approximately only one in five.

Moreover, there is yet another problem. Assuming one formulates the function correctly, one not only has to think of the function. One must *also* think of the application of the function. That is, one has to think of the items constituting the speech context *as argument* and then assign the value to that argument. ". . .one has to think of the member of *D* one calls 'this'. If one conceives of the application as an inferential move, then one has to think of the members of *D* and assign the value in question. *But then one is already thinking of the object. How does one think of the object*?"[25]

In sum, probabilistic and possibilistic measures cannot capture the semantics of indexicals such as 'this' and 'that' in ordinary, natural language discourse, let alone those indexicals that are non-linguistic such as images. We are left with no account whatsoever of the use of even public natural language indexicals let alone the primitive relations of immediate awareness and the indexical reference found there. We have no account for the fact that one is already thinking of the object which shows that the mind is already in an immediate relation with that object.

6.11. Summary

For purposes of argument, we set aside the issue of whether or not neural networks can handle *sui generis* objects of primitive relations found in immediate awareness, to consider whether or not, and to what extent, these computer architectures can simulate kinds of knowing found in Boundary Set *S*. The Self-Organizing Feature Map is the best candidate and was the only unsupervised and self-organizing architecture considered. However, we found that it is not successful in extracting or organizing hierarchical categories from streams of sensory data. As noted, such streams are often highly cluttered and noisy, containing a lot of irrelevant information that must be "selected out" to permit kinds of natural intelligence exhibited in the cocktail party problem.

We also evaluated whether or not neural networks can handle ordinary linguistic indexicals. Utilizing a common, ordinary indexical usage found in natural language, we found that neural network computer architectures cannot handle even ordinary linguistic, context-dependent indexicals, let alone non-linguistic ones. The basic obstacle stems from the fact that indexicals cannot be encoded as mathematical functions. To try to encode them as such would minimally result in begging the question, as I argued above.

Finally, I should add another problem for computational approaches to immediate awareness and knowing how. A feature of all computer architectures, including artificial neural networks, is the simple fact that they all adhere to a basic input-output model. That basic model may be multiplied many times over, stitched together to form a more complex set, and functionally designed to run as a distributed, parallel process. Nonetheless, they still adhere to a simple input-output model that is limited in its information-theoretic properties. It does not permit information-theoretic extensions into the environment that in turn permit system interactions and *transactions* with objects in whatever constitutes the environment. Kinds of knowing found in Boundary Set *S* are exemplars of natural intelligence in interaction and transaction with objects, specifically *sui generis* objects, in the environment and within the person.

---

1 Simon Haykin, 1994.

2 Simon Haykin, 1994.

3 Becker, 1991; also Haykin, 1994.

4 The graceful degrading of the network is also ensured by "coarse coding" which I have not discussed here. The idea is that features [particulars] are in effect spread over several neurons, ensuring that the parallel distributed processing approximates the flexibility of a continuous system rather than the rigidity of discrete symbolic AI systems. [See Hinton, 1981, 1986]

5 Adam Blum, 1992.

6 There are many measures for determining similarity, which I will not thoroughly discuss here. However, a commonly used measure is based upon the concept of Euclidean distance defined by the difference between a pair of *N*-by-1 real-valued vectors all of whose elements are real. [See Haykin, 1994]

7 I rely heavily upon Simon Haykin's *Neural Networks, A Comprehensive Foundation* [1994] throughout.

8 Simon Haykin, 1994, p. 397.

9 Though not entirely. Again, an epistemological theoretical approach is broader in scope than a neurophysiological one: for instance, it includes the primitive relation of imagining which we cannot map with a histogram.

10 Simon Haykin, 1994.

11 Lotfi Zadeh, 1981.

12 All references to Zadeh's Test Score Semantics will be to his 1981 source.

13 Zadeh, 1981 p. 283.

14 Zadeh, 1981 p. 283.

15 Hector Neri Castañeda, 1981, p. 281.

16 With some modifications, this example is in part patterned after an example given in Castañeda, 1990, p. 741.

17 Lotfi Zadeh, 1981, p. 289.

18 Zadeh, 1981, p. 289.

19 As Zadeh notes, *t* may also be interpreted in truth-functional semantics as the truth value of *p* given D, *i.e.* $t = \text{Tr}\ \{p \mid D\}$; also, in possible world semantics, *t* may be interpreted as the possibility of D given *p*, *i.e.* $t = \text{Poss}\ \{D \mid p\}$.

20 For more discussion of these particular assumptions underlying set-theoretical approaches to semantics, see Castañeda, 1989, pp. 140-144.

21 Castañeda, p. 140 ff.

22 Castañeda, 1989, p. 141.

23 Ibid.

24 Castañeda, 1989, p. 143.

[25] Ibid., p. 144.

7. COMPUTABILITY OF BOUNDARY SET *S*

"...This non-computational process lies in whatever it is that allows us to become directly aware of something." Roger Penrose[1]

In the above chapters, I have set forth a view of immediate awareness and knowing how as highly complex, self-organizing and adaptive categories of human knowing. The bases for these categories were found in ontological as well as epistemological analyses and arguments, carrying forth a tradition established by James and Russell. In particular, I have characterized *knowing the unique* as a multifaceted set consisting of a hierarchy of primitive epistemic relations of immediate awareness, and suggested that we must look to a broader theory of signs, within an even broader theory of indexicality, to fully understand it. In this last section, I want to consider more abstractly the computability, that is decidability (in principle), of Boundary Set *S*.

In order for a computer program to solve any abstract problem, problem instances must be represented in a way that the program can understand. An encoding of a set of abstract objects is a mapping *e* from *S'* to the set of binary strings [or any set of strings over a finite alphabet having at least two symbols; however, the complexity of the problem may vary with "expensive" alphabets]. Obviously, we can encode the natural numbers N = {0,1,2,3...} as the strings {0,1,10,11,100,...}; we also have the EBCDIC and ASCII codes.

On classical computability theory, an abstract problem is a binary relation on a set *I* of problem instances and a set *S'* of problem solutions. That is, classical computability theory is limited to decision problems having "yes" and "no" answers to questions of the form "Does there exist a solution to. . .?" However, this is not the appropriate form for epistemic problems involving *knowing how* and *knowing the unique*. Enactive signs such as gestures, patterns of actions making up the performances of tasks, and other actions such as touching and moving, *all of which are significant in natural intelligence,* require mappings with vector notation, the computable reals or complex numbers. It is clear that the machines [or algorithms] over the complex emergent dynamics of *knowing how*, especially where *knowing the unique* is embedded within it, must minimally be machines over the reals or complex numbers.[2] A computer can generate highly complex emergent dynamic behavior by, for example, successive iterations of mappings of computable complex numbers onto an Argand plane. The evidence for that complex emergent dynamics is found in the Mandelbrot and Julia sets[3] produced by such iterations.

An encoding of a set of abstract objects is a mapping *e* from *S'* to the set of binary strings [or any set of strings over a finite alphabet having at least two symbols; however, the complexity of the problem may vary with "expensive" alphabets]. Graphs, mathematical functions, ordered pairs, polygons, and programs can all be encoded as binary strings. I earlier displayed a graph of the universe of signs by which the universe of knowing, the epistemic universe, is representable, exhibited, or disclosed, that is *signed.* The category of signs included symbolic, iconic, and enactive signs. Viewed as an encodable universe, however, there are differences between the kinds of signs and there are problems with encoding certain ones.

Each category of sign is encodable. Symbolic and iconic signs are easily encodable into binary strings. However, enactive signs such as gestures require vector notation, the computable reals or complex numbers. Researchers[4] at California Institute of Technology have built a gesture recognition system by the following means. They used filmed gestures encoded as sequences of frames with each frame containing a number of pixels. A pixel representation permits considering a gesture as a sequence of arrays. Then by stacking the arrays, a three-dimensional space with time can be created. Motion characteristics were incorporated in the three-dimensional space, followed by the application of pattern classification techniques. But their system is clearly not intentional gesturing or intentional recognition.

Minimally, enactive signs require the rationals, however this would still not be sufficient to precisely characterize kinds of *knowing how*.[5] Thus, robotics, a domain of computation theoretically isomorphic with the domain of *knowing how* in its control architecture, must encode its algorithms with the reals or complex numbers [or minimally, though deficiently, with the rationals].

To the extent that our *knowing* or *epistemic universe* must be defined over the reals, *R*, or complex numbers, *C*, as implied above, classical recursive function theory must be extended to handle such questions of decidability, a matter I cannot address in depth here. Such an extension entails that concepts such as *recursive enumerability*, R.E., also be redefined. Moreover, *rule-governedness* is usually defined as an R.E. set, under the classical definition of R.E. We want to keep our classical understanding of *rule-governedness*, but we must also extend our concept of rules related to immense as well as uncountable sets to include *rule-boundedness.*[6] Rule-boundedness can characterize some sets over immense domains as well as the uncountable domains of the reals and complex numbers. Again, though we cannot explore further these extensions and redefinitions of concepts of classical recursive function theory, such extensions will permit a broader epistemological theory construction and concept of *computability* related to a broader concept of human intelligence based upon categories of *knowing how* and *knowing the unique*.[7]

The human being [not limited to the human brain, but including the entire human body] can generate in its behavior highly complex, self-organizing, emergent dynamics in the performance of apparently very simple tasks. We can mathematically characterize this behavior with computer generated images by iterated mappings. To understand problems related to the computability of our Boundary Set *S*, must review some concepts and problems of computation. One of my overall aims throughout this book has been to support arguments that a rightly understood epistemological universe must include in its domain the real or complex numbers and hence cannot be directly handled by the classical discrete computational approach to problems defined over its domain. And, unlike the classical formalisms developed by Gödel, Turing, Church, Post, *et al.*, resulting in an identical class of computable functions [(partial) recursive functions], we do not have a machine [algorithm] that will take as input any element of the R^2 or C universe and output 1 or 0 depending upon membership in the set.[8] Though there exists a variety of *ad hoc* approaches to this and related problems, an invariant theory [set of formalisms] has not been developed.

Moreover, related to the problem of finding an effective procedure over the reals, R^2, or C, are the complexity issues, which I also cannot address here. Extant complexity theory is also based upon the models and formalisms of classical computation theory in which complexity problems are defined over the integers [or mechanisms which can be effectively encoded into the integers], hence the theory does not extend to problems defined over the reals or complex numbers.

7.1. Computation and Complex Epistemic Domains: Problems with the Classical Computational Approach to Boundary Set *S*

As stated above, the formalisms of classical complexity theory are founded upon the models and formalisms of classical computability [decidability] theory. The complexity of a machine or algorithm [on the classical model] is the maximum number of steps, that is elementary machine operations for solution over all inputs or problem instances of size L. A function or problem is tractable [solvable in polynomial time] if it is computable by a machine with complexity function bounded above by a polynomial function of L. Otherwise, the function or problem is intractable. The concepts of computability, decidability, and complexity are all defined over the integers or mechanisms encodable into the integers.

However, there are many computational problems in such areas as robotics, geometry, and epistemic [knowing] problems merged within these, such as those found in Boundary Set *S*, *knowing how* and *knowing the unique,* which have as their natural domains the reals or complex numbers. If we define an epistemic set or domain exhibiting complex dynamics, such as a hyperspace with trajectories of kinds of epistemic vectors, we would view the universe [of the problem space] as C, the complex numbers, or as the real plane R^2. The question then becomes: is there a machine that will take as input any element of that universe and output 0 or 1 depending upon membership in that epistemic set?

Under classical computability theory, there is no algorithm which will in general take these elements as input. If the epistemic universe is a universe over the reals or complex numbers, as I have argued *knowing the unique* and kinds of *knowing how* are, then we cannot formulate our decidability/computability questions about that universe following classical theory. For example, as noted above, the terms 'recursive' and 'recursively enumerable' classically apply only to countable domains. If our epistemic universe is uncountable, how can we make sense of the question "Is the epistemic set *recursive*?" Moreover, how would a new theory of decidability and a new theory of complexity, extended to account for domains over the reals and complex numbers affect the complexity classes of languages, the P, NP and NPC?

As earlier noted, the *epistemic* universe is a set exhibiting very complex dynamic, self-organizing behavior. Recall that the epistemic universe consists in the following :

$$\text{Knowing} = \text{Subject} \cup \text{Object} \cup \text{Content} \cup \text{Context}$$

This universe as a whole has a very complex geometric structure with the set of points making up its boundary exhibiting a very rich and extraordinarily dynamic and complex structure, on analogy with the Mandelbrot and Julia sets.[9] Kinds of knowing found there are very much as Kauffman has described as "poised on the edge of chaos." The latter sets are used here to illustrate certain properties of the epistemic universe as a whole and the boundary set in particular to show the serious limitations on the classical computational/decidability approach defined over the integers.

The universe of knowing is *in part* a very large population of simpler components, machines. In the development over time of a human knower those simpler components are not primarily *knowledge that* but are primarily simple components of knowing found in the intersection of *knowing the unique* and *knowing how*. For *knowing the unique*, the epistemic primitives are the "species" of that knowing I identified earlier. This population of simple machines, over time, constructs aggregates of simple rule-bound epistemic objects which interact and *transact* nonlinearly with one another and their environment to produce emergent epistemic structures of knowing. The system is a complex, dynamic, self-organizing, and emergent system.

7.2. The Decidability of the Epistemic Boundary Set *S*: Issues From the Moral Universe

To approach the computability of epistemic Boundary Set *S* as represented by kinds of *knowing how*, such as knowing how to probe an open wound, or knowing how to be kind to others, we must look carefully at the problem space and (briefly) at the nature of the complexity measure on it. We might approach the problem as the following task: "Produce a decision procedure for the above set *S*." We must view the epistemic sets *knowing the unique* and *knowing how* and their intersection in *S* as sets of vectors, as an ordered set or list of variables, characterizing multivariate computations. The vector components are real numbers, coordinates of points in a hyperspace. Again, each vector will be a member of the set of all vectors [*arbitrary* real numbers] denoted by R^2. For now, my intention is to be intuitive, to get at the core of problems with formalisms over the reals. What I refer to as the epistemic or knowing state of a person can be characterized by state vectors which is a space consisting of all combinations of values of the epistemic variables in the ordered set which defines the space.

The classical computability theorists wanted to understand the concept 'decidability' or *decidable set* in order to make sense of such questions, "Is the set of theorems of arithmetic decidable?" Again, a set *S* in a universe of discourse *U* is decidable if there is an effective procedure that, given any element *u* of *U*, will decide in a finite number of steps whether or not *u* is in *S*. That is, it will decide if the characteristic function of *S* relative to *U* is *effectively computable*. We are concerned with the set *S* and wish to decide, for any element *u* in *U*, whether or not *u* is in *S*.

To address the decidability of Boundary Set *S*, it may be helpful to diverge somewhat from my prior practice of focusing upon single concrete human performances to illustrate very abstract matters, in order to present some of these same very abstract theoretical computational issues within a very human context. This may help us to better highlight those "things that go to make a human being" that Dreyfus was concerned to show could not be computerized. That context is the current, very human debate within philosophy between those who are identified as *virtue ethicists* and *rule ethicists.* Ethics generally and the study of moral theory in particular has always been fraught with difficulties leading some to reject the basis of such concerns in reason altogether. I certainly will not endeavor to present all the issues and various positions here in assessing decidability of Boundary Set *S*, but will center on issues dividing the two camps surrounding the nature of ethical theory itself.

Some rule ethicists argue that in many respects virtue ethics is best interpreted as *anti*-theory, that it constitutes a philosophical brief against all efforts to set forth yet another formal theory. Virtue ethicists argue that formal theories overly simplify extremely complex matters in human life by always leaving out something, and are virtually useless in coming up with universal rules which apply in all cases of moral and ethical conflict at all times. For the other side, this smacks of relativism leading to no ethics at all. Both camps appear to agree on one thing: being ethical and moral has a lot to do with one's doings, that is, with one's knowing *how, where, when, what* and *in what right proportion* to do, or *not* do, something. The moral universe is a place filled with the complexities of knowing --or not knowing how.

Given the extended epistemological theory and rather abstract distinctions presented so far on rule-governedness and rule-boundedness, I want to propose another way of viewing (not resolving) the issues. I want to especially consider those criticisms raised by virtue ethicists against the formal nature of ethical theories. This is in fact an ancient contest between those who champion on each side the old categories of speculative reason and practical (knowing how) reason. On one side of the debate, I believe virtue ethicists are largely correct to claim that modern formal ethical theories do not accurately reflect the dynamic complexity of actual moral practices, reasoning, and concepts. But the response to this fact, I believe, is not to reject theory or adopt quasi-theoretical, non-formal, and largely relativist approaches as virtue ethicists appear to be doing as they grope toward a different conception of what ethical theory ought to look like. I argue that what is needed is a theoretical formal approach more suited to the continuous (*not* discrete), dynamic, and vague nature of the actual moral universe. This may not be helpful in one's day to day deliberations, but it may help us to understand ourselves better than we do.

As with classical computation theory, modern ethical theory also largely relies on a model of theorizing formalized in the 1930's. The bases for the formalisms were around a long time before then and are apparent in the approaches to ethical theorizing and decision procedures long before (and since) they were developed in the 1930's by Gödel, Turing, Church, and Kleene. In essence, as noted above, it is an approach which defines decision, "yes" and "no" problems over the natural numbers $N = \{0, 1, 2, 3. .\}$. Relative to a universe or domain U, for example the moral universe, a set is decidable if there is some effective procedure for deciding for any given element u in U, whether or not u is in that set.

Virtue ethicists are largely on target, I believe, with their criticisms of this (now) classical formal decidability approach to ethical theorizing. The history of modern ethical and moral philosophy shows that the nature of the actual moral universe (set) and its elements (moral concepts, rules, practices) have been defined narrowly, apparently to suit the discrete (computational) formal theoretic approach at hand, rather than developing the continuous, dynamic theoretic approach necessary. It is a classic case of allowing the "tools" at hand to determine the problem, rather than the reality of the problem determining the development of tools necessary to grapple with it.

To resolve the issues, some theorists argued for a radical intuitionism, premised upon the following assumptions: (1) the moral universe is unruly; (2) there are no sound rules for combining component moral considerations into a single comprehensive imperative or measure of value. Because of these, (3) practical moral reasoning becomes "reasoning by analogy" from an unruly universe to a *rule-governed* universe, and this practical reasoning includes at least three kinds of components: (a) a set of rules, which *though unsound* in the [unruly] universe as it stands, governs a related rule-governed [moral] universe; (b) a set of rules for constructing and drawing conclusions from analogies between the two universes; and (c) a set of cases, derived in no rule-governed way, but derived from intuition or experience, or maintained as ideals or paradigms.[10]

This is a *risk averting* model, where the risk to be avoided is that of reaching unsound conclusions, because (it is claimed) we cannot be sure what kind of universe we inhabit. And, (even stronger) "we cannot *know* what kind of [moral] universe we live in." That is, we cannot know if our moral universe is: (i) recursive; (ii) rule-governed but not recursive; (iii) not rule-governed but not unruly, that is partially rule-governed; or (iv) unruly. Therefore "[w]e can minimize the risk of reaching unsound conclusions by adopting a method of practical reasoning that applies even in an unruly moral universe."[11]

The first thing to notice is the correlation of *rule-governedness* with *recursive enumerability,* without a concept of *rule-boundedness*. That is, a rule-governed set must have an effective procedure for listing (counting) its members. Additionally, there is an implicit assumption that to be corrigible, a moral universe must be (or must be treated as) a countable set. These assumptions are natural to those who assume a computational theory of mind, the view that human beings are at best Turing machines, precise algorithms for decision processes. If the computational theory of mind is wrong, one holding such a position would conclude that ". . .we seem to be left with unanalyzable, mysterious abilities that can't be explained."[12]

Of course, everything I have argued for thus far in this book leads me to the view that this position presents us with a patently false dichotomy, a dichotomy which says that the moral universe is either comprehensible by means of computer algorithms or it is incomprehensible, ultimately "mysterious."

In addition to examining what might be a place for the concept *rule-boundedness* in this debate, we might explore what practical moral "reasoning by analogy" might amount to *on analogy with* reasoning by analogy regarding our [physical] universe. The above notion of practical moral reasoning by analogy is a risk aversion model because "we cannot be certain what kind of moral universe we inhabit."

But we might ask at least the following question: Why assume that the direction of *certainty* is the direction of *moral* being in that universe? Perhaps the state of being less likely to be in error about our moral conclusions and decisions is an index of the *narrowness* or *simplicity* of the moral questions and arguments we pose to ourselves. Such narrowness or simplicity may lead to incalculable harm to others.

Moreover, averting risk might be an index of our efforts to *avoid* actual moral complexity and genuinely perplexing moral problems. Yet again, it might be an index of the poverty of our conceptual methodological and theoretical tools to address the moral complexity of the problems actually found there. If so, avoiding or neglecting these problems would seem to have the effect of reducing our development--over time--as moral persons. The above proposal to avoid risk in moral decision-making would seem to be equivalent to avoiding moral complexity. Hence, the "risk aversion" model, in effect, provides us with a severely truncated view of the nature of moral reasoning.

Reasoning by analogy regarding the *physical* universe, on the other hand, is actually a *high* risk venture (in which mistakes are often made), but such ventures and mistakes are often fruitful in shedding more light on the physical universe we inhabit. This is the reason why we say that a failed experiment is better than no experiment at all. Though we do not seek to fail, a failure nonetheless gives information for deciding on experimentation in the future, and may end up revealing much about the universe that we did not know before.

There are substantial differences between reasoning by analogy regarding the physical universe and the practical reasoning by analogy regarding the moral universe in the above proposal. These differences are too numerous to be explored here, but I suggest the following comparison as the core of my concerns with the use of the classical computational approach: In many respects, the use of what amounts to the conceptual tools of classical computational theory and formal logical systems to address the nature of and to resolve moral problems is similar to attempts to use Euclidean geometry to describe the shape of a cloud, a mountain, or a tree. Clouds are not spheres, mountains are not cones, and bark is not smooth.[13] Such an effort is bound to fail or at best provides us with a distorted view of what is really there in the natural world. The patterns of Nature and the patterns of a moral universe are both irregular and fragmented compared to the precise lines of Euclidean geometry--or the lines of an argument and proof in formal systems. Both Nature and the moral universe exhibit not only a higher degree but *an altogether different level* of complexity. The number of considerations in a moral universe, as well as the number of distinct scales of length of natural patterns, are for all practical purposes at least immense, and cannot be accounted for with the classical computational approach applied to practical moral reasoning.

For these reasons, I propose that we seek alternative strategies for addressing the nature and complexity of the moral universe in an extended domain of epistemology which includes immediate awareness, knowing the unique, and knowing how, and in complex dynamical systems theory, with a concept of *rule-boundedness*.

In rejecting the above concept of *rule-governedness* which is on analogy, appropriately enough, with a straight-edge ruler, I propose we view the phenomena found in a moral universe more on analogy with, for example, the actions of Brownian motion. This may strike some as odd, but I will argue that the intense and often fragmented, irregular trajectories of lines of moral considerations and reasoning about them have more in common with these irregular patterns of nature than with a straight-edge ruler---or with the lines in a classical formal proof of any kind. Given this, I also propose an alternative view of the kinds of human knowing involved in moral considerations, moral actions, and moral reasoning. Those kinds of knowing are found in Boundary Set *S*.

7.3. Kinds of Knowing Found in the Moral Universe

Clearly, classical as well as contemporary rule ethicists tend to emphasize *knowing that* moral proposition and rule as a necessary condition to moral decision or action. It is an obvious emphasis on propositions rather than on persons. These theorists assume what Ryle referred to as the "intellectualist legend" to explain the move from knowing a moral proposition or rule to a moral action. Practical moral reasoning, knowing how to be moral, is simply to "do a bit of theory, then do a bit of practice." *Knowing how* to be kind, to be courageous, to be honest and loving is explained largely in just those terms, though the "fit" of the explanation is not entirely comfortable. And in such theories, the immediate awareness, the felt sensitivity of persons which permits one to single out unique qualities of persons, events, and objects is generally not recognized at all.

Yet in a moral universe, to know a person *qua person,* as the unique individual they are, is to know them as uniques, not as mere members of classes or lists of their properties. Such knowing of a person is not reducible to, nor derivable from, knowing propositions about them. But again, the emphasis in the tradition is on propositions, not persons.

Though I disagree with assumption (1) above, that the moral universe is unruly, based on the correlation of recursive enumerability with rule-governedness, I think it is quite correct to assume (2), that there are no sound rules for combining moral considerations into a single comprehensive imperative or measure of value. But there is an ambiguity in claim (2) which seems to blur a necessary distinction between rules and those moral considerations to which they apply. That is, this appears to make both of these elements of the universe which the writer claims is unruly. Moreover, it may very well be correct that the universe (or set) of sound rules for combining moral considerations into a single value or comprehensive imperative may be null, but it is not clear whether the set of moral considerations itself is more adequately characterized as recursively enumerable but not recursive, or whether it should be characterized as not countable at all. In some respects which I will explore below, it appears that it is not countable.

Though the assumption above is of a non-sound-rule-governed, that is unruly universe of rules for combining moral considerations into a single value or imperative, there is also an assumption that there are real *moral* considerations (as distinct from *non-*, *a-*, or *immoral* considerations) which are components or elements which make up a set. The rules that are missing from the moral universe are just those that permit combining these elements (moral considerations) into a single imperative or measure of value. So the set, where the elements are such rules, is not recursively enumerable. That is, we don't have an algorithm for generating the rules for combining moral considerations into a single value.

7.4. Recursively Enumerable But Non-Recursive Moral Sets: Is the Set of Moral Considerations a Countable Set?

However, if there is a moral universe where this refers to a set of moral considerations, where the elements are just these considerations, *that* set might be generated by some algorithm even if there are no sound rules for combining them into a single value or imperative. I tentatively take 'moral consideration' to refer to those descriptions of "the morally relevant aspects of situations and actions." In short, I will refer to these as moral propositions which might be, in some sense, subject to proof. If there is such an algorithm, such a set would be recursively enumerable, but it does not follow that both that set and its complement are recursively enumerable. That is, we may have a set which is recursively enumerable but not recursive. But if there is no such algorithm, we may have a set to which the terms 'recursive' and 'recursively enumerable' do not apply, that is we may have a set that is not countable.

We might consider what is implied by this for a moral universe (set) and its consequences to rule-governedness as well as rule-boundedness. I will consider sets (finite or infinite) of natural numbers. Each single natural number *n* is taken to represent a proposition which is a moral consideration. The set of all moral propositions of the formal system are represented as the entire set *S*. The theorems of the formal system, if there are any, might constitute a smaller set of natural numbers, for example, the set *R*. To set up a one-to-one correspondence between the natural numbers and the moral propositions, Q_n, we need a known algorithm for obtaining each Q_n from its corresponding natural number and another one for obtaining *n* from Q_n.

If we're fortunate in our universe of moral considerations, at least some of the moral propositions $Q_1, Q_2, Q_3 \ldots Q_n$ in our set will have proofs, and they will constitute some set in *S*, that is the set *R*. Thus, if we have an algorithm for generating the elements of *R*, we have a recursively enumerable set. We will also have the set of moral propositions whose negations are provable, and that set is recursively enumerable as well. But it can be proved that *R* is recursively enumerable but not recursive. That is, the complement of *R* is not recursively enumerable. Moreover, the formal system is not complete because there are moral propositions which are neither provable nor disprovable in the formal system. Those propositions are undecidable.

It may be worthwhile to briefly review the argument why this is so, with help from Penrose.[14] If we assume our formal system includes the actions of all Turing machines,[15] we can denote the *n*th Turing machine by T*n*. For each natural number *n* in our formal system we can express the following proposition:

(1) 'T*n* (*n*) stops'

The set of all such propositions (1) as *n* runs through the natural numbers will represent some subset of *S*, call it K. (1) will be true for some values of *n* and false for others. Turing showed that there is no algorithm that asserts 'T*n* (*n*) does not stop' in those cases in which T*n* (*n*) in fact does not stop.[16] That is, he showed that the set of false (1) is not recursively enumerable. If we had an algorithm for generating the elements of the complement of *R,* we would be able to enumerate each (1) in that complementary set. These would be the false (1). But since the false (1) are not recursively enumerable, the set *R* is non-recursive. Additionally, the set of false 'T*n*(*n*)' is incomplete, that is we have a set of undecidable propositions which are neither provable nor disprovable in the formal system. Moreover, the subset of true moral propositions of *S* is not recursive nor recursively enumerable, nor is the complement of that subset recursively enumerable.

Viewed geometrically, the implications of all the above to our moral universe are interesting. A recursive moral set would be one with a relatively simple, clear and clean-cut boundary. It would be a relatively easy matter to decide, for any given moral consideration, whether or not it belongs to the set or its complement. It is quite a different matter with recursively enumerable but non-recursive sets. The boundaries of such sets or geometric figures are extremely complex, like the boundary of a Mandelbrot set.[17] As with moral considerations in our day to day situations, there are sometimes "morally dense" regions in which it is very clear which considerations are there and which ones are not. There are "blobs" of dense points in the set. But there are also curling and meandering tendrils, floating from the dark regions of the set with increasing subtlety toward the light regions of its complement. All with a curious sameness or self-similarity nonetheless. It often isn't clear where and to what degree a "moral point" belongs to the set or doesn't.

Figure *SEVEN-1.* The Mandelbrot with Julia Sets

For non-recursive sets, there is no algorithmic way of deciding whether or not an element belongs to the set. We do not have a general algorithm for deciding whether or not an element belongs to the set. Moreover, as Penrose notes, it isn't *certain* that the Mandelbrot set is recursively enumerable. It seems, however, that its complement might be, that is it (the complement) may be a recursively enumerable set which is non-recursive. Perhaps in many situations, we're better at deciding what *isn't* morally relevant to a given situation than what *is*.

With respect to the latter, we are often left very unsure, unless we limit our decisions to the "morally dense" regions represented as blobs in the Mandelbrot set. That would be similar to limiting our moral decisions to a narrow and simplified selection of moral considerations and arguments which we present to ourselves. That is, it would be to avoid the complexity that is actually there. But that is often to miss the moral significance of a given situation entirely. It is sometimes the subtlety of the actual meandering tendrils that tell the moral tale.

And it is the set of those actual meandering tendrils of a complex set such as the Mandelbrot that "go off into infinity" which present real problems for classical computational theory and any formal ethical or moral theory based or modeled upon that classical theory. Whether one assumes a consequentialist (teleological) or non-consequentialist (deontological) ethics, or any of the theories falling under each category, a pervasive underlying assumption is that one *can* sort moral considerations out from all other considerations, though each position may differ as to what they claim those considerations are, and may attribute different values to them. That is, a pervasive assumption is that the moral set--whatever each position takes that to be--is a *decidable* set.

As stated above, relative to a universe *U*, a set is decidable if there is some effective procedure that, given any element *u* of *U*, will decide whether or not *u* is in the set. That is, a set is decidable if its characteristic function, the function defined on *U* that has value 1 on the set and value 0 off the set, is a *computable* function. This means that the question of whether or not a set is decidable *is* the question "What is a computable function?" and it is natural to then ask if moral reasoning is computable. That is, can a machine reason morally?

But the classical theory of computation runs into trouble with complex sets such as the Mandelbrot set. A complete explanation is too complicated to be pursued in extensive detail here, however, the primary reasons the classical computational theory fails are: (1) that theory is fundamentally a *discrete* theory with mappings over the domain of natural numbers (permitting bivalent "yes" and "no" answers), while moral phenomena are fundamentally *continuous and complex dynamic* (and hence would require mappings over at least immense domains or the reals--and answers may be found to span an infinite unit interval); and (2) the complexity of sets such as the Mandelbrot render them effectively *undecidable* (on a classical computational approach).

These considerations also suggest that rule-governedness of the moral universe, correlated with computability, cannot be equivalent to coming to *know* or having learned a moral concept completely. First of all, not all concepts are given. Human beings also *form, create,* and *discover* new concepts. That is, we form, create and discover ones which were either not existing or not known beforehand. At one time, the concept of proof for Fermat's last theorem was not completely known, but was nonetheless discovered (evidently) by Wiles.[18] If ethics or the moral universe is similar to mathematics in this respect, there are moral concepts we have yet to discover, create, or form and they may not be *rule-governed* in the sense of being recursively enumerable. We may not be able to list, that is count, all instances of that concept.

Moreover, of those concepts that are given, which one might correctly consider necessary to ethical and moral language, for example, the concept of courage, not all are perfect in the sense of having clear boundaries. The imperfections in our concepts, or the *naturally occurring* irregular boundaries, may lead to errors in our reasoning. All the more so if instead of attending to the means to improve our given concepts, or improve our reasoning about them, we merely relearn their imperfections or the imperfections in our reasoning from one generation to the next.

As stated above, I agree with the above assumption that there are no sound rules for combining moral considerations into a single value or comprehensive imperative. Justification for our moral choices cannot always proceed *directly* by appeals to rules at all (formal or informal). And the above position advocates a form of reasoning by analogy to provide an indirect method of appealing to rules to seek moral justification. At this point, we might review one or two pivotal assumptions on which this position and approach is based.

If justifiable moral reasoning must proceed on the basis of rules *in some sense* (and I certainly agree that it must), then there is much to support the above argument. But I believe it is missing much of the greater complexity of a real moral universe, a complexity which requires taking another look at our understanding of *rule-governedness* and looking seriously at the concept *rule-boundedness* as it pertains to our own knowing and understanding. As seen above, a rule-governed set is a countable set. We find rule-governedness *only* in sets we can count. This also means that the concept 'rule' must include at least the following: (1) it is expressible in well-formed linguistic strings; or (2) it must be translatable into a well-formed linguistic string. Rules are rules only if they can be expressed in (countable) language (such as English or some alphanumeric artificial language) propositions or imperatives. But this leaves out immediate awareness and non-linguistic indexicals altogether. The above position does not consider rule-boundedness where this may only be *exhibited* in some non-linguistic way, and is not formulizable, or countable, even in principle.

On the other hand, if we extend the notion of rule-governedness to refer to that which is *non-random* in linguistic and other behavior (of natural and human systems), we must take into consideration other ways besides language that human beings have of exhibiting or representing what they know. We must look in the moral universe not only for knowing *that* proposition or rule, and knowing *that* such and such consideration is morally relevant to the situation, we must also look to kinds of knowing *how* and knowing *the uniqueness* of persons, events, objects.

I suggest that we can make better sense of the moral universe, of the distinction (and false dichotomy) between virtue and rule ethics, by turning to an broader epistemological point of view and complex dynamical systems theory. If our epistemological categories are adequate, they will provide a more comprehensive way of viewing the relations between virtues. They will also assist us in better understanding the legitimate criticisms raised against an exclusively rule-based ethics as well as those raised against an exclusively virtue-based ethics.

In most parts of the moral universe, no matter how well- or how ill-equipped we are with rules of the formulizable sort, *knowing how, when, where*, and *in what right proportion* to correctly apply or employ them is beyond the domain of any mechanical countable procedure. And that is only the beginning of our problems. There is also a problem of knowing unique qualities of persons, events, and objects, sufficient to provide the sensitivity or intimacy necessary for awareness of the moral complexity of a situation, a discernment of the *incomparable* features like no other which are morally relevant. *Knowing the unique* is not reducible to, nor is it derivable from knowing a formulizable rule or language proposition. It is not reducible to *knowing that*.

7.5. The Epistemic Universe as Complex Numbers, *C*, or the Real Plane, R^2 and the Undecidability of Epistemic Boundary Set *S*

If we assume the epistemic universe *U* is *C*, the complex numbers, or as the real plane R^2 [with z = x+iy in *C* with (x,y) in R^2], that is not just the computable but arbitrary reals and complex numbers, then we are asking if there is a machine that will take as input any element *u* of *U* and output 1 or 0 depending on membership of *u* in *S*, our boundary set. Again, on the classical theory, there is no such machine, though there are *ad hoc* models and methods used to analyze computational and complexity problems in this area, including interior point methods and the "real number" approach.

Generally, interior point methods such as the Karmarkar algorithm are rational number model approaches. Blum [1989] illustrates the problem with these using the linear programming problem (LPP) in which one is looking for a linear function on a polytope in R^2 defined by inequalities $Ax \leq b$, $x \geq 0$ [where 'A' is an $m \times n$ matrix and $b \in R^m$]. What one looks for is a highest point, generally a vertex. The Karmarkar and other interior point methods start at an interior point and take a series of discrete steps or iterates, converging to a highest point. However, interior point methods generally do not halt at a solution, though the process can be stopped in a prescribed number of steps at a good or approximate solution. An exact solution is then calculated from the approximation.

As Blum explains, the Karmarkar algorithm is presented and analyzed as the "rational number approach." She explains that the underlying justification for the approach, supportive of the use of models and paradigms of the discrete classical theory of computation, appears to be the following:

(1) Machines are finite.
(2) Finite approximations of the reals are the rationals.
(3) Therefore, we are really looking at problems over the rationals.

Also, the Simplex algorithm, presented as the "real number model" approach, is finite but exponential in the worst case. For example, using the above linear programming problem, for each *n* (dimension parameter), there are instances of the LPP of dimension order *n* [even with small integer coefficients] that take around 2^n pivots for solution.

Thus, Blum concludes that both competing algorithms for a problem falling within the interface of the discrete and continuous theories, are defined and analyzed using incomparable approaches: the Simplex algorithm is finite but exponential in the worst case in the "real number" model, while the Karmarkar algorithm is polynomial (P) in the "rational number" model, but not even finite in the "real number" model. In sum, there is no genuine real number formalism or model, and those *ad hoc* approaches to analysis and computation of such problems either do not halt at precise solutions or they are exponential in the worst case. Where computability [decidability] and complexity theories are extended to the reals and complex numbers, an open problem would be "Is the LPP polynomial in the [genuine] real number model? Again, however, at this point we do not have a formalism defining the real number model.

On the classical theory, and strictly within the domain of *knowledge that*, Gödel[19] showed that given any reasonable, that is consistent and effective theory of arithmetic, there are true assertions about the natural numbers that are not theorems in that theory. That is, any reasonable theory of arithmetic is necessarily incomplete because for any such theory there will be arithmetic sentences whose truth or falsity cannot be decided on the basis of the axioms and rules of inference of the theory. Obviously, the concept of undecidability is crucial to understanding Gödel's work.

On the other hand, if we direct our question "Is the set *S* decidable?" to our set *S* as defined above, there is a problem already with the question itself, since the very concept of decidability is defined over the natural numbers [or mechanisms such as gödel coding sentences encodable in the integers].[20] Gödel's Incompleteness Theorem and classical computability theory in general, have as their domains the natural numbers *N*, extended to the integers *Z*, the rational numbers Q, or mechanisms encodable in *N*.

Though Gödel's Incompleteness Theorem is a theorem derived from axioms over the natural numbers in which he essentially asserted a true but not provable proposition within the system of *Principia Mathematica*, much interest has been focused since then upon the issue of how pervasive or common the incompleteness and unprovable phenomenon is. With respect to the epistemic universe, this issue really comes down to the question "Can the knowing universe be comprehended?" Included in this question of course is the question "Can mathematical knowing be comprehended?" With respect to the latter, Hilbert believed that every mathematical question could be answered. Any mathematical assertion could be shown to be either true or false by means of a mathematical proof.

Of course, Gödel showed that Hilbert's belief was mistaken. With respect to our broader question "Can the knowing universe be comprehended?" we already know by Gödel's proof that it cannot be completely comprehended by means of a formal proof, where that part of the knowing universe is defined over the integers. Moreover, even where we define other parts of that universe over the reals, there are further arguments and proofs within information theory and complexity theory[21] which show that there is information-theoretic incompleteness even there. That issue is beyond to objectives of this study, though it can be argued that there is additional evidence beyond Gödel of the fundamental undecidability and incompleteness of Boundary Set *S*.

7.6. Summary

I have evaluated the very abstract question, "Is Boundary Set *S* decidable?" Where immediate awareness and knowing how are definable over the reals or complex numbers, as I have argued they must be, then we cannot formulate the decidability question itself following the classical theory. To unfold more of the issues involved, I considered the moral universe and some debates between rule and virtue ethicists.

The history of modern ethical and moral theory shows that the nature of the actual moral universe (set) and its elements (moral concepts, rules, practices) have been defined narrowly, apparently to suit the discrete (computational) formal theoretic approach at hand, rather than developing the continuous, dynamic theoretic approach necessary. Rule ethicists have not recognized the rule-bounded nature of many ethical and moral issues, adhering solely to rule-governedness. There is an implicit assumption by the rule ethicists that to be corrigible, a moral universe must be (or must be treated as) a countable set. Otherwise, it is unanalyzable and mysterious. These assumptions are natural to those who assume a computational theory of mind.

However, I argued that the history of these problems showed that these arguments are a classic case of allowing the "tools" at hand to determine the problem, rather than the reality of the problem determining the choice and development of tools necessary to grapple with it. I also argued that what is needed is a theoretical formal approach more suited to the continuous (*not* discrete), dynamic, and vague nature of the actual moral universe.

Based largely on Gödel's incompleteness and undecidability

theorems, with some considerations from Chaitin applications to information theory, Boundary Set *S* must be considered undecidable.

---

[1] Roger Penrose, 1994 [emphasis mine], p. 53.

[2]The computational domain of performative knowing how intersecting with knowing the unique is coextensive with the control architectures found in some robotics. However, it should be made clear that the indexical *use* of a sign in a relation of presentation is problematic with respect to its encodability and realizability by a computer. The translation of the indexical use into a computable language changes the epistemic status of such a sign to the representational relation.

[3]The Julia set is the set of points of a Mandelbrot set that do not go off to infinity under iterations of mappings of a complex number onto a plane or sphere. The *filled* Julia set can be formed from a polynomial map of the form $g(z) = z^2 + c$ on the Riemann sphere $S = C \cup \{\infty\}$.The Mandelbrot and Julia sets are products of iterations of nonlinear transformations [mappings] of the real line. The universe of the set is *C*, the complex numbers, or the real plane R^2 [with $z = x + iy$ in *C*; (x,y) in R^2]. Whatever complex number we choose, that number is represented as some point on the plane. With each iterated mapping, g, z is replaced by a new complex number given by $z^2 + c$ [with c another given complex number; on Blum's filled Julia set, $c = -(.39059 + .58679i)$]. This is then also represented as a new point on the plane. More iterations follow, with particular numbers replaced by new complex numbers, c, and the complex numbers themselves replaced by $c^2 + c$. From this procedure, we obtain a sequence of complex numbers mapping points on a plane. With certain sequences of complex numbers, we obtain an array of points on the plane that are never far from the point of origin. The sequence remains within a fixed circle (boundary of radius) centered at the origin. That is, the sequence remains bounded for those choices of c. Moreover, as Penrose has proposed, following classical decidability theory, we can ask "Are the Mandelbrot and Julia sets decidable sets?" The question is somewhat awkward or inappropriate because those sets are not generated by algorithms defined over countable domains as the classical decidability theory is so defined.

[4] Dinkar Gupta, "Computer Gesture Recognition: Using the Constellation Method," in *Caltech Undergraduate Research Journal*, Vol. 1, April 2001.

[5]As noted above, classical computability theory demands that we define problems or function instances over *N*, the natural numbers, *Z*, the integers, or *Q*, the rationals, or mechanisms encodable into *N*, a *countable* domain. As Blum points out [1989], extending beyond *N*, problems arise when we are tempted, as is prevalent in much current literature, to consider only rational *Q* skeletons of problem sets, thus confining our focus to points on a rational grid, Q^2. There are "gaps" or "holes" in a rational graph of such problems. For example, a real graph of $x^3 + y^3 = 1$ in the positive quadrant will look like a quadrant of a circle, which might be one of our complex epistemic sets. However, the corresponding set of rational points on the graph will be empty [See Blum, 1989; also Lay, 1990].

[6]Certainly, as I have made apparent throughout, knowledge *that* is a set defined over rule-governed, that is R.E., sets. Such sets are the only ones classical epistemology has dealt with, and are the only ones most philosophers currently wish to deal with.

[7]I explore the extensions of classical recursive function theory and its relation to epistemological theory construction and a broader concept of intelligence in my 1995 (in progress). To understand the nature of Boundary Set *S*, I have utilized the mathematical characterizations of the Mandelbrot and Julia sets, within the broader context of Boolean network theory. The dynamics of our set *S* become clear by referring to the properties of the Mandelbrot and Julia sets, which also assist in highlighting problems with extant

computability theory to address decidability questions on those sets as well as our set *S*. There are a number of questions which a complete theory must address, which we are unable to address here. These include: (1) What is the nature of rule-governedness and rule-boundedness on that epistemic set *S*? More to the point, how are we to use the mathematical characterization of *rule-boundedness* to make sense of *epistemic* rule-boundedness? This question entails the following questions: (2) If our epistemic boundary set *S* is *like* the filled Julia set in significant mathematical and epistemological respects, how can we use the Julia set to mathematically characterize and understand our own knowing? (3) How do we encode primitive elements of the presentation set, immediate awareness (knowing the unique), and levels of this primitive epistemic set, including *touching imagining* and *moving*, to get the dynamics we need to understand our own knowing? (4) How do we mathematically define *manner of a performance* which is central to the dynamics of our set *S*? and (5) Since classical computability (decidability) theory is limited to machines (algorithms) over discrete, *countable* domains, and our boundary set *S* is definable over continuous, *uncountable* domains, how do we extend classical recursive function theory to make sense of questions regarding the decidability of our set *S*? Though I cannot address these questions here, we can at least consider the properties of the Mandelbrot and Julia sets and see that they have a usefulness in a mathematical model of natural knowing systems, specifically a mathematical characterization of our boundary set *S*.

[8]The set of all real numbers is denoted by *R*. Tools for arithmetical work on the reals are the operations of addition, multiplication and relations between reals such as equality (=), 'greater than' (>), and 'less than' (<). Components of vectors will be the real numbers; each such vector will be a member of the set of all vectors, like (a,b) where a and b are arbitrary real numbers. The set of all vectors is denoted by R^2.

[9]For the sake of argument, I assume the epistemic universe is like the *filled* Julia set of a polynomial map on a Riemann sphere where $S=C \cup \{\infty\}$ of the form $g(z) = z^2 + c$. The boundary Julia set is the set of points that don't go off to infinity under iterations of g. [See Blum, 1989].

[10] Bonevac, Daniel, "Ethical Impressionism: A Response to Braybrooke," in *Social Theory and Practice,* Volume 17, no. 2, Summer, 1991, pp. 157-173.

[11] Ibid.

[12] Ibid.

[13] Benoit B. Mandelbrot, *The Fractal Geometry of Nature*, New York: W. H. Freeman and Company, 1977.

[14] Roger Penrose, *The Emperor's New Mind: Concerning Computers, Minds, and the Laws of Physics,* New York, Oxford: Oxford University Press, 1989.

[15]A.M. Turing, "On Computable Numbers, With An Application to the *Entscheidungsproblem*," in *Proceedings of the London Mathematical Society*, Volume 42, 1937, pp. 230-265.

[16]Penrose, Ibid., pp. 121-122.

[17]See Penrose, Ibid., pp. 74-79.

[18] See Simon Singh, *Fermat's Enigma: The Epic Quest to Solve the World's Greatest Mathematical Problem,* New York: Walker & Co., 1997

[19] Kurt Gödel, "Über Formal Unentscheidbare Sätze der Principia Mathematica und verwandter Systeme I, " in *Monatshefte für Mathematik und Physik*, vol. 38, 1931, pp. 173-198.

[20] Lenore Blum and S. Smale (1990). *The Gödel Incompleteness Theorem and Decidability Over a Ring*, Technical Report, Berkeley, California: International Computer Science Institute.

[21] G. J. Chaitin, "Information Theoretical Limitations of Formal Systems," *Journal of the Association of Computing Machinery*, Volume 21, 1974, pp. 403-424

8. SUMMARY AND CONCLUSIONS

Immediate awareness is our primitive, nonlinguistic, nonpropositional *knowing* in our experience in the world. It is that richly textured tapestry of deeply embedded primitive relations of our sensory and somatosensory-motor system with which Nature has endowed us, enabling our very survival. It is the most fundamental and pervasively embodied cognitive network underlying all our natural intelligence.

I have argued for a broader theory of knowing, a broader theory of natural intelligence. I have also showed that elements of our knowing are, in principle, not computable. Following in the realist tradition of Bertrand Russell and the pragmatic tradition of William James, I have given meaning to the concept 'immediate awareness' as a set of primitive epistemic relations of *knowing*. These primitive relations are not found in language representations of any kind, but in the living person. I also showed how those primitive relations include the multilayered primitive relations of *touching* and *moving* to provide an account of bodily kinaesthetic intelligence. This is intended to overcome the Cartesianism of earlier views. Moreover, my theory has been presented with a broader theory of *sign* relations, not limited to alphanumeric *symbolic* or linguistic relations, and is a more complete epistemological classification than found in earlier theories.

My theory of immediate awareness is based primarily upon arguments showing that sensation is not cognitively neutral. Our sensory and somatosensory-motor processes are not representational processes, and we slip into subtle nominalist fallacies when we take our language metaphors too far. Immediate awareness is not mediated by propositional, linguistic maps. It is not a set of beliefs nor is it based upon belief. Nonetheless immediate awareness is a kind of knowing. It is the most primitive cognitive network underlying all our natural intelligence. Deeply embedded within us, it permits as well as drives our *knowing how*, our bodily intelligence underlying all other kinds of our intelligence.

My theory is based on the patent observation that, contrary to a prevailing view, sensation, sensory and motor activity are *not* cognitively neutral, but a map of the most fundamental layer of our multiple, natural intelligence, kinaesthetic-bodily intelligence. This map underlies and is interwoven with all other cognitive, intelligent activity of any kind.

8.1. What the Facts of Natural Intelligence Show

The facts of human knowing and intelligence show:

1. Current theories of natural intelligence and human knowing cannot account for even the most basic things human beings *know how* to do.
2. Current theories of intelligence are based on a narrow epistemology that claims or assumes that *knowledge that,* based on kinds of belief, is the only kind of human knowing.
3. Theories of natural intelligence reflect this narrow epistemology by their assumption that intelligence is a single entity and can be measured with verbal "paper and pencil" tests.
4. Current theories of natural intelligence and human knowing are also based largely on unstated Cartesian assumptions that there is a fundamental split between body and mind.
5. The most pervasive Cartesian assumption underlying current theories of intelligence and knowing is the "intellectualist legend," that *knowing how* is always the application of a *knowledge that* rule or prescription. To know how to do anything at all is to "do a little theory" then "do a little practice."

6. Current popular Naturalist theories of intelligence and knowledge are largely nominalist and materialist.
7. Among other things, nominalism and materialism confuse a symbol for the thing symbolized, leading to the collapse of levels of inquiry and wholesale fallacies based on the collapse, as well as illegitimate reductionism and begging the question (among others).
8. Naturalist theories have failed to provide an adequate methodology to address the comprehensive scope of human knowing, thus also fail to address the comprehensive scope of natural intelligence.
9. These spurious theories of natural intelligence and human knowing have carried over into Artificial Intelligence and Artificial Life theories.

Contrary to these views, early arguments and empirical evidence showed that *knowledge that* and *knowing how* name two distinct, non-reducible kinds of intelligence. My arguments and empirical evidence presented here show that cognitive immediate awareness is embedded within *knowing how*. It is exhibited in the seamlessly smooth ways in which people do even ordinary, every day tasks. Immediate awareness, knowing the unique, is not reducible to kinaesthetic-bodily intelligence, in part due to immediate awareness of abstract objects. It is also not reducible to *knowledge that.* In fact, empirical evidence shows that knowing the unique often correlates negatively with verbal reports.

Also contrary to the above, I have shown there are three major, nonreducible categories of human knowing, of human natural intelligence: (1) knowledge that; (2) knowing how; (3) knowing the unique (immediate awareness). Both empirical evidence and logical argument show that human beings are in fact endowed with multiple, nonreducible kinds of natural intelligence that are interrelated in highly complex, dynamic and self-organizing ways we do not understand.

I also presented extensive empirical evidence and argument showing that cognition begins not with the attention phase, but much earlier, minimally during the preattentive phase; it is not mapped isomorphically with language use, but is much broader and deeper.

Both empirical findings and logical argument lead to the conclusion that we must carve the space of human knowing, cognition, of natural intelligence, much differently from the prevailing view. We must continue and further research efforts on the preattentive structures, and the sensory and somatosensory-motor systems by which human natural intelligence exhibits that it knows more than it can say. We also need to develop theoretical means by which to map that natural intelligence to computers so as to extend our understanding even further. We need to look more closely at the human use of indexicals, such as intentional gestures and other movements to point to our knowing, a fact not accounted for in traditional and most prevailing views of human intelligence. We already know that indexicals are not limited to alphanumeric *symbolic* or linguistic relations. What we need is a more complete theory of indexicality and a broader classification of indexicals than found in current theories.

8.2. Themes

My arguments established a more geometric view of the universe of natural intelligence, away from classical symbolic, "top-down" views biased by Cartesianism and the propositional "intellectualist legend" decried by Ryle. A more geometric view permits the expansion of that universe to include the self-organizing, dynamic complexity of actual human knowing, including knowing how and immediate awareness.

A more geometric view of the universe of natural intelligence requires the distinction between *rule-governed* and *rule-bound* knowing. This distinction is necessary (though obviously not sufficient) to account for what Dreyfus referred to as *commonsense know how and understanding of human beings*. It is a distinction which is at once mathematical, computational, epistemological and neurophysical, intended to pave the way toward resolution of fundamental problems in computational approaches to immediate awareness.

My arguments show that the most promising approach to the computability of immediate awareness is a "weak AI" position involving the use of random Boolean networks and complex dynamical systems theory. The use of random Boolean networks permits us to obtain some of the fundamental properties of self-organization of autocatalytic sets.

I have argued that immediate awareness may be conceived as a kind of autocatalytic set of dynamic primitive relations made publicly manifest in primitive relations of *knowing how*. The use of random Boolean networks shows a way of obtaining *law-like properties* of those primitive relations of immediate awareness in terms of dynamical systems theory without committing one to a physicalist/materialist epistemology. This is in contrast to Penrose's quantum-mechanical Objective Reduction strategy, which I believe entails a fundamental materialist fallacy. It gives us a way of understanding core properties of our own inner conscious lives, and of understanding the smoothly timed and seamless sensitivity of primitive somatosensory-motor awareness. This direction for a theory of knowing [broader than just a theory of *knowledge about*] was implicit in the work of both James and Russell, though they did not have the concept of nonlinear function, and Russell (as opposed to James) was wedded to an atomistic, summative ontology.

I have taken issue with the standard "strong AI" position on the computability of immediate awareness by reassessing some of the very basic *philosophic* as well as computational and neurological concepts upon which our understanding of intelligence [or knowing] largely rests.

Much of the current debate on the issue of consciousness is pervaded by prior largely hidden Cartesian and reductionist assumptions. One of the most pervasive of these is the assumption that all knowing (or cognition generally) is reducible to propositional know*ledge*, representable in declarative sentences (or encodable as such), based upon beliefs. Thus knowledge representation theories abound.

What Dreyfus earlier referred to as commonsense know how and understanding of human beings is to be found in the intersection of *knowing how* and *knowing the unique.* I called that intersecting set *Boundary Set S.* It includes kinds of knowing which are clearly not propositional and not *rule-governed* though I argued that they are *rule-bound.*

I also argued that there are serious limitations on the computability of human knowing, especially those kinds of knowing found in Boundary Set *S*. All such computational models to date, as applied to immediate awareness, are either not self-organizing, while immediate awareness clearly is, or if they are self-organizing (as Kohonen's map), they are unable to accommodate the clearly hierarchical nature of the sensory and somatosensory *cum* motor structure of the brain. But even if they could, they would still miss *sui generis* objects of thought, of immediate awareness, altogether. This is so because meaning representation languages for encoding natural language expressions conflate grammatical meaning with mathematical functions. They cannot handle even linguistic indexicals, let alone non-linguistic ones actually found in *knowing how*.

Moreover, in both the computational and neurophysiological research, including recurrent, multilayer neural network theory, models of classification processes of *concept* formation are often taken to be adequate to account for *percept* formation of immediate awareness, when they are not. Again, classification processes of computation cannot handle unique, *sui generis*, objects of immediate awareness (which is why I call it *knowing the unique*).

On the basis of purely formal arguments, I argued that Boundary Set *S* is not decidable, thus not computable on the standard von Neumann computer.

8.3. Comments on Some Contrasting Views

My views are clearly at cross-purposes with some other recent writers on this or related topics, such as Block [1995] and Chalmers [1996]. A clear strength of Block's work is his effort to establish a *nonrepresentational* concept of what he calls *phenomenal* consciousness. His concept of phenomenal consciousness is not equivalent with my concept of *immediate awareness* though they both share certain properties. However, I believe his overall effort suffers largely due to his nominalist *cum* psychologistic strategy to address what is fundamentally an epistemological problem, though he does not recognize it as such. His strategy fits with Quine's earlier program to shift all epistemological questions to psychology.

But because of this, Block fails to get at the fundamental nature of consciousness generally, particularly *phenomenal* consciousness, due to certain unstated and unexamined nominalist myths about the nature of human cognition and reason. Among those myths is a very narrow construal of cognition itself. As other nominalists, Block holds that all cognition or intentionality must be representational, having "contents" representable in *that*-clauses. They must be representable in *that*-clauses, he says, in order to be accessible to reasoning. This is merely a restatement of what Ryle earlier called the "intellectualist legend" at the core of Cartesianism. It is the assertion that all intelligence is mind-based *knowledge that*, and that intelligent performance or action must be preceded by an intellectual acknowledgment of *knowledge that* rules or criteria, the content of *that*-clauses.

Of course the watershed contribution Ryle made to philosophy was to show that this view amounted to a philosopher's myth and is fundamentally mistaken. But Block builds an entire set of distinctions between what he calls "phenomenal consciousness" and "access consciousness" on this fundamentally mistaken view of intentionality and cognition. For Block, it is as though Gilbert Ryle never existed. For some of these same reasons, one finds the Cartesian "intellectualist legend" in Block's underlying view. On his view, in spite of both empirical and theoretical evidence to the contrary, there is no such thing as bodily kinaesthetic intelligence.

Furthermore, at the core of much of Block's analysis are the implied nominalist reductions of a *symbol or code of* something with *the something symbolized or encoded.* Consistent with his largely nominalist strategy, he does not address the underlying *ontological* dimensions of his inquiry at all, and does not see the consequences for his own aims of not doing so. I should make clear that I agree with Block that there are fundamental distinctions to be made between what he calls *phenomenal consciousness* and *access consciousness,* though I do not believe he has successfully made any such distinction, and I do not believe he has adequately conceptualized the nature of either kind.

Calling the most intractable kind of consciousness *phenomenal,* already begs numerous questions regarding the nature of the *objects* of that consciousness as well as the means of being conscious of them. The term 'phenomenal' refers to objects of the senses, that is things one is conscious of through the senses, as opposed to objects of thought or [what Penrose calls *intuition*]. Though Block's and Chalmers' terminology is consistent with current trends, the older *realist* tradition found in the works of James and Russell referred to an expanded construal of this kind of consciousness as *immediate awareness,* for some very good reasons that I have set out in chapters of this book. Again, however, the term 'immediate' was not originally intended to mean "meaningless," but to emphasize a kind of "oneness" with the object(s) of which one is aware, without a language or representational interface. But both Block and Chalmers give rather short shrift to the concept *awareness*, evidently unaware (no pun intended) of its previous philosophical significance and relation to the object of their concerns.

If one takes a scientific approach to consciousness seriously, then one must take a more *realist* as opposed to *nominalist* view of the issues. A realist view requires an ontological analysis, though, again, it is precisely that analysis which is almost totally missing in most current publications in consciousness studies.

Some ontological and other assumptions and issues underlying Block's as well as Chalmers' views, but nowhere addressed by either of them, include questions of the sort that even Gödel[1] pointed to as crucial to understand our own understanding and awareness. Those questions include: "What is a *thing*, an ordinary *object* of any kind?" "Are all things or objects of any kind merely *classes* or members of classes of some kind?" and (equivalently) "Are all objects or things *extended*, that is merely sums or lists of their properties or predicates?" If consciousness is a function, how does this differ from consciousness as a relation? Paraphrasing James, "Are we witnessing today the usurpation of metaphysics by language?" in the "strong AI" and nominalist presumption that all thought is computable? Are there objects which can only be *present* but not *represented* in our experience?" "Indeed, what is *experience* itself?"

Much of the current literature on consciousness is shot through with all kinds of assumed answers to these and other kinds of questions, though with certain exceptions the questions themselves have not been asked at all. More egregiously, one finds an uncritical acceptance of some prevailing unanalyzed notions regarding the nature of experience as such. The domain of experience is demarcated in such a way as to not only disengage it from much of its traditional meaning, as largely sketched out in Russell's theory of knowledge, but to also rule out certain kinds of questions we might ask about experience and its relation to kinds of consciousness. For example, one sees the phrase "what it is like" used to refer to experience.

For some theorists, "what it is like" is supposed to distinguish *phenomenal* from *access* consciousness. Yet close analysis shows this to be a bogus distinction, that "what it is like"actually refers to a description or proposition, statable in a "that" clause. For that very reason, however, it cannot be what distinguishes the two. And while Block notes that there is a sense in which phenomenal consciousness can only be pointed at, criticizing others [*e.g.* Searle] for pointing to too much, he does not offer an outline of an account of [intelligent] pointing itself. Indeed, on his assumed equivalencies between cognition, intentionality and that which is represented in "that" clauses [tied to that which is encodable in computer programs], there are no such things as acts of *rational* pointing. Again, I believe all this is due to a nominalist influence found in many current publications on consciousness.

With respect to *The Conscious Mind*, Oxford, 1996, more than just about any other contemporary philosopher, David Chalmers has publicly insisted on forcing the "hard problem" of consciousness into the public arena. He has provided a beneficial service to the academic and research communities by insisting that they pay serious attention to this problem. However, I find much the same nominalist problems with Chalmers' effort that I found in Block's work, though again there is very clearly a sincere desire to understand the nature of this most difficult problem. On the positive side, Chalmers insists on taking the problem seriously, and says he insists on taking a *science* of consciousness seriously. But, as with Block's effort, the nominalism *cum* psychologism clearly evident in his work is at cross purposes with this apparently sincere aim to take consciousness seriously in a scientific sense. We find evidence of the cross purposes very early in his book, in his discussion of efforts to define consciousness:

> Trying to define conscious experience in terms of more primitive notions is fruitless. One might as well try to define *matter* or *space* in terms of something more fundamental.[2]

But *matter* and *space ARE* defined, scientifically, in terms of "something more fundamental." Their meanings are explicated by appealing to additional, undefined or primitive terms. 'Matter' is defined in terms of *mass* [as anything that has mass]; 'mass' is further defined in terms of the measure of inertial and gravitational properties; and 'inertial mass' is defined through Newton's second law, $F = ma$, and so on. 'Space' is actually 'space-time', the central concept of the theory of relativity, mathematically formulated by Lorenz prior to the interpretation by Einstein, with an abstract description by Minkowski. In short, what one gets with a *serious scientific approach* to admittedly very difficult concepts, is *a very rigorous, precise mathematical approach in which primitive terms are taken seriously indeed.* Of course, definition is not all there is to doing science. But definitional chains leading to laws and law-like descriptions, which are the *summum bonum* of science, are crucial. Primitive, undefined terms are essential to those chains.

What we get in the initial chapter of Chalmer's book, which essentially sets the stage for the entire book, is a largely anecdotal discussion of conscious experiences with largely facile, *ad hoc* distinctions drawn. I do not find any fundamental, *logically drawn*, theoretical framework for a rigorous study of consciousness. Like Block, Chalmers has a narrow view of the scope of intentionality because he ties it to *belief*, represented in "that clauses." This is also due, perhaps, because he has failed to see the serious significance of Ryle's work, especially to his own endeavors. Though Chalmers mentions Ryle, he says nothing about Ryle's basic watershed contribution to epistemology.

Most problematic is Chalmers' discussion of and arguments for a "Strong AI" approach to consciousness. He draws an *ad hoc* distinction between internal and external objections to machine consciousness, stating that external objections such as those raised by Dreyfus, "have been difficult to carry through, given the success of computational simulation of physical processes in general."[3] I had to read this section over several times to make certain I was not misunderstanding him. Chalmers must not aware that there is *extremely limited* success of any kind whatsoever in computational simulations of precisely the sort which was of concern to Dreyfus. And this is so because--setting aside the computational simulation issue--we do not yet understand *very basic processes* involved in body movement, that is, whole body displacement. It should be noted that our scientific knowledge and understanding of human moving [movement generally, or whole-body displacement] is quite limited. We do not as yet even understand how moving is stored in the memory, how we spatially image or reconstruct a trajectory path [path integration] in our minds, or how we "home" in on a target, objective or goal with our bodily movements.[4] These matters appear to depend upon a kind of context sensitive awareness, of which we have little understanding. It is that context sensitive awareness that I have attempted to provide for in my theory of immediate awareness.

8.4. Conclusion

I have endeavored to provide a partial theoretical account of immediate awareness and its relation to knowing how, with a bridge over the Cartesian gap between rule-governed *knowledge that* and rule-bound *knowing how.* I do not believe that Dreyfus' challenge to Strong AI has ever been answered, though I do believe many have come to see the significance of his arguments in their magnitude and complexity. I for one hope to help pave a theoretical path toward their solution.

Of necessity, this effort remains incomplete, given the enormous research and inquiry yet to be done. If there is a single proposal I wish to make after this long journey, it is that we must look again at how we have carved the space of natural intelligence. We must also set about to re-carve it, based in part upon evidence and arguments I have presented here. How we understand that space is ultimately how we understand ourselves and our place in the universe.

Myrna Estep

---

[1] Kurt Gödel, "What is Cantor's Continuum Problem?", in *Philosophy of Mathematics, Selected Readings*, Paul Benacerraf and Hilary Putnam, (eds.), New Jersey: Prentice-Hall, 1964.

[2] David Chalmers, *The Conscious Mind*, New York, Oxford: Oxford University Press, 1996, p. 4.

[3] Ibid., p. 313.

[4] See Berthoz, Alain, Isabelle Israël, Pierre Georges-François, Renato Grasso, Toshihiro Tsuzuku, "Spatial Memory of Body Linear Displacement: What is Being Stored?" *Science*, American Association for the Advancement of Science, Volume 269, 1995, 7 July, pp. 95-98.

APPENDIX

Proper Names and Definite Descriptions: The Sense and No-Sense Theories[1]

Usually, the distinction between proper names and definite descriptions is drawn in order to determine, in part, how words or language generally relate to the world or reality. Moreover, the term 'language' is usually delimited to alphanumeric symbol systems of one sort or another. These include both natural languages such as English or German, and formal languages such as mathematics and computer languages. It is usually the case that 'language' is narrowed to a class of symbols or sentences about reality and rules governing their use. The concept is not usually taken to include physical gestures or motions with the body, including eyes and hands, as well as intonation when speaking (though one will sometimes find some consideration of intonation), unless rules for these are themselves statable in the language. If we include physical gestures as part of what a language is, then we might broaden the definition of 'language' to *sign* systems, which include symbols, as opposed to narrowing them to symbol systems. One can "sign" meaning with words or with the use of gestures as signs to point to an object of thought and reality.

The term 'proper name' is usually likewise delimited to words or alphanumeric symbols which one can write down or speak, though at least some philosophers, notably, Bertrand Russell, held a broader and more abstract view of the concept of proper name.[2] He took the concept of a *logically* proper name to be a primitive relation to an object of thought in acquaintance or immediate awareness. Logically proper names essentially *point* to or individuate the object; they do not describe the object in any way. Moreover, logically proper names do not point to universals, but to particulars. Unlike universals, the same particular is only *in principle* accessible to more than one person; as a matter of empirical fact, however, the same particular is rarely if ever experienced by more than one person. Obviously, for Russell logically proper names are not those proper names one usually thinks of and finds in natural language, such as 'John', 'von Clausewitz', and 'London'. The latter are taken by him to be elliptical for or disguised definite descriptions.

Historically, the issue regarding proper names and definite descriptions has been taken to be the following problems: First of all, what are proper names? Do they differ from descriptions? If so, how do they differ?

In general, though there are differences of philosophic opinion on the matter, a common view of proper names is that they are ordinary names found in any natural language. Thus ordinary names such as "Maria," "David,"and "Chicago," are held to be genuine proper names. A commonly held view of definite descriptions is that they are "the reflection on the window," "the guy sitting next to John," that is phrases or labels which include lists of properties which uniquely describe objects.

The No-Sense Theory of Proper Names and Frege's Sense-Theory Response

One philosophic position argued for is that proper names simply stand for objects they name. This is the *no-sense* theory of proper names. The phrase 'no-sense' is used because though proper names are taken to denote objects, they have no connotation or sense. A connotation or sense is a description. Proper names on this theory are held to differ from common nouns like 'cat' which connote a set of properties which can be set forth in a definition of the class, and the common noun also denotes the class of cats. Thus, unlike proper names, common nouns have both connotation as well as denotation. In other words, a proper name does not describe the object it names, it is not a class name or label. In one possible version of the no-sense theory, proper names function to point to the object, though the theorists usually holding this view have not generally referred to that function when describing the theory. At most, they have implied an indexical, individuating, or ostensive function of proper names. In any case, proper names are distinguished from descriptions on this no-sense theory.

However, Gottlob Frege[3] noted that if proper names simply stand for, that is denote, objects and nothing more, then we are left with the question: How do identity statements, including proper names, ever convey factual information? For example, one sees proper names in identity statements such as "a is identical to b". If such statements are only about the referent of the names, then they are trivial. If they give information about the names, then they are arbitrary since we can assign any name to an object. Frege argued that besides the names and the objects they refer to we have to distinguish the sense or connotation of the name in virtue of which it refers to an object. In the sentence "The evening star is identical with the morning star", "the evening star" and "the morning star" have the same referent but different senses. The sense provides the different mode of presentation of the object. What the statement conveys is that one and the same object has different senses of the two names and has two different sets of properties specified by the two different senses of the two names. Thus such a statement is a statement of fact and not a mere triviality or an arbitrary verbal decision. All proper names for Frege have senses in the way that the expressions "the evening star" and "the morning star" have senses.

But one might argue that Frege has missed the issue of *genuine* proper names altogether in the following ways: (1) He is speaking solely of *language about* proper names and the objects of [not even genuine] proper names. He is clearly addressing solely those names found in natural languages; (2) But the names found in natural languages are not genuine proper names. Genuine proper names are not ordinary names like 'evening star' and 'morning star' because the existence of their objects is a contingent fact and in no way follows from the status of the expressions in the language. These are in fact disguised definite descriptions. Thus, (3) Frege is in fact addressing descriptions, not proper names. But one might keep in mind William James'[4] lament that language has taken over our metaphysics. As found in Frege, if we do not have a linguistic name for an object, there is therefore no object. Naming objects is naming reality, and the obverse of this is that if we have no name [description] for an object, there is no object. Thus reality is isomorphic solely with language. Moreover, an object is no more than a list of its properties which is its name, that is its description.

The classical no-sense theory held that genuine proper names *necessarily* have a reference but no sense [connotation or description] at all. But on Frege's view they have a sense and only *contingently* have a reference. They refer if and only if there is an object which satisfies their sense [connotation or description]. Of course, this reduces proper names to definite descriptions. On the classical theory, however, proper names are *sui generis*. The concept *sui generis* means "its own kind, or one of a kind, unique". But we might argue that even this is in fact misleading since to call such an object "one of a kind," is to assign a description to it. To speak of *kind* is to speak of a description, a class. But such objects *qua sui generis* are not "of a *kind*". That is, there is no *kind* of which it is one. An object which is *sui generis* is entirely unique *like no other* regardless of any properties or predicates it may share with any other object or kind. Thus no matter how many properties it may share with others of some kind or class, the object of a genuine proper name is unclassifiable. For Plato [*Theaetetus*] and Wittgenstein[5] [*Tractatus*] they are the special connecting link between words and world. For Frege and others who follow a sense theory of proper names, proper names are simply definite descriptions. That is, they are class, not unique, objects.

As Searle remarks in his essay on proper names and descriptions,[6] common sense inclines us toward the no-sense theory when we are speaking of ordinary names found in natural language. That is, proper names such as 'John' and 'Mary' or 'San Francisco' do not seem to be definite descriptions because when we ordinarily call an object by its name we are not describing it. Also, we do not have definitions or their equivalents for most proper names. Moreover, a name is not "true of" its bearer, it is its name. Not only do we not have definitional equivalents for proper names, but it is not evident how we could get such definitions.

But it is claimed that the difficulties for the no-sense theory remain: (1) It cannot account for the occurrence of proper names in informative identity statements. Since I have argued that these do not contain genuine proper names in the first place, this is no objection to the theory. (2) It is unable to account for the occurrence of proper names in existential statements. The same argument given above applies because these do not contain genuine proper names. Moreover, this is shown by Frege's own position on existence: it is a second-order concept. An affirmative existential statement does not refer to an object and state that it exists; rather it expresses a concept and states that that concept is instantiated. Thus, as Searle rightly points out, if a [genuine] proper name occurs in an existential statement, it must have some conceptual or descriptive content. However, genuine proper names do not have descriptive content, and they do not occur in existential statements. Indeed, as Russell makes clear,[7] they do not occur in statements at all.

However, the no-sense theory still faces the following difficulty, as raised by Searle: (3) What account can the no-sense theorist give of the existence of the object referred to by a proper name? In the *Tractatus*, Wittgenstein held that the meaning of a proper name is literally the object for which it stands. He later stated that this was not correct because he had confused the *bearer* of the name with the *meaning* of the name.[8] But Searle's reply to Wittgenstein's earlier stance on the problem, though partially correct, betrays the usurpation of metaphysics by language earlier lamented in the quote from James:

> If one agrees with the Wittgenstein of the *Tractatus*. . .then it seems that the existence of those objects which are named by genuine proper names cannot be an ordinary contingent fact. The reason for this is that such changes in the world as the destruction of some object cannot destroy the meaning of words, because any change in the world must still be describable in words.[9]

We might agree that the existence of objects referred to by genuine proper names cannot be ordinary contingent facts, while disagreeing that "any change in the world must still be describable in words". But Searle is confusing definite descriptions with proper names. The object of a genuine proper name is not describable in words, and it is those objects of which Wittgenstein speaks in the *Tractatus*, whether speaking of the meaning or the bearer of the genuine proper name. Searle's reply misses the point because he does not recognize the limits to language, especially the limits of description.

While it is true that the objects (or bearers) of genuine proper names are not ordinary contingent facts, since they would then be characterized by propositions, it does not follow, as Searle wants to then argue, that they are therefore a class of objects whose existence is somehow necessary. Searle is using criteria for evaluating the ontological status of these objects which are appropriate for logical objects, the objects of classes or sets. But the objects of genuine proper names are *sui generis*. They are not class objects.

But historically, according to Searle, there have been two alternative ways offered out of the problem of the existence of these objects. These include a metaphysical way taken by Wittgenstein in the *Tractatus* and a linguistic way expounded upon by Anscombe[10] in *An Introduction to Wittgenstein's Tractatus*. It is my position that Wittgenstein's early metaphysical path is the more philosophically sound one, given corrections on it that he himself recognized needed to be made. Anscombe's linguistic path is one which has led to the kind of nominalism *cum* idealism much in evidence today. It misses the point of genuine proper names altogether. It results in an unacceptable relativism which outstrips metaphysics with linguisticism by fiat. This has led to the absurdities noted above, earlier recognized by James in his work on the relation of knowing in *Essays in Radical Empiricism* [1912].

Searle's[11] own proposed solution is actually no solution at all. He addresses the use of names in natural language which are labels or elliptical for definite descriptions, noting that the imprecision of such names in natural language is an effective linguistic convenience. Proper names for Searle are nothing more than "pegs on which to hang descriptions." While this may very well be true of names found in natural language, it completely misses an understanding of the *sui generis* nature of genuine proper names altogether and is certainly no solution to the existence of their objects.

---

[1]For a short history of this problem, I have relied upon Searle, John, "Proper Names and Descriptions," *Encyclopedia of Philosophy*, Paul Edwards (ed.), Volume 6, New York, Macmillan Publishing Company, 1967.

[2] Bertrand Russell, "The Philosophy of Logical Atomism," in *Logic and Knowledge*, R.C. Marsh, (ed.), London, 1956, pp. 200-201, but especially in Russell's *Theory of Knowledge: The 1913 Manuscript*, Elizabeth Ramsden Eames, (ed.), London and New York: Allen & Unwin, 1984.

[3] Frege, Gottlob, "Sense and Reference," in P.T. Geach and Max Black, (eds.), *Translations From the Philosophical Writings of Gottlob Frege*, New York, Oxford: Oxford University Press, 1952.

[4] William James, *Essays in Radical Empiricism*, Longmans, 1912.

[5] Ludwig Wittgenstein, *Tractatus Logico-philosophicus*, translated by C.K. Ogden, London, 1922.

[6]John Searle, “Proper Names and Descriptions,” in *The Encyclopedia of Philosophy*, Vol. 6, New York, Macmillan Publishing Company, 1967, pp. 487-491.

[7]Bertrand Russell, *Theory of Knowledge: The 1913 Manuscript*, Elizabeth Ramsden Eames, ed., Allen & Unwin, London and New York, 1984.

[8]Ludwig Wittgenstein, *Philosophical Investigations*, translated by G.E.M. Anscombe, Oxford, 1953, paragraphs 40-79.

[9]Searle, “Proper Names and Descriptions,” in *The Encyclopedia of Philosophy*, Vol. 6, p.488.

[10] Anscombe, G.E.M., *An Introduction to Wittgenstein's Tractatus*, London: Hutchinson University Library, 1959.

[11]Searle, Ibid.

REFERENCES

Acar, W. (1988). "Theory Versus Model, Further Comments," *Systems Research*, Vol. 5, 172-173.

Adleman, Leonard (1995). "A Boom in Plans For DNA Computing," *Science*, American Association for the Advancement of Science, Vol. 268, 28 April, pp. 498-499.

Albus, James S. (1981). *Brains, Behavior, and Robotics*, Peterborough, New Hampshire: BYTE.

Albus, James S. (1991). "Outline for a Theory of Intelligence," in *IEEE Transactions on Systems, Man and Cybernetics*, Vol. 21, No. 3, May/June.

Allgood, K., J. Yorke (1989). "Fractal Basin Boundaries and Chaotic Attractors," in *Proceedings of Symposia in Applied Mathematics: Chaos and Fractals*, Volume 39, American Mathematical Society.

Almog, Joseph, John Perry, Howard Wettstein, (eds.), (1989). *Themes From Kaplan*, New York, Oxford: Oxford University Press, 1989.

American Heritage College Dictionary (1993). Third Edition, Boston, New York: Houghton Mifflin Company.

Anderson, Adam K., and Elizabeth A. Phelps (2001). "Lesions of the human amygdala impair enhanced perception of emotionally salient events," in *Nature*, Vol. 411, 17 May, pp. 305-309.

Anscombe, G.E.M. (1959). *An Introduction to Wittgenstein's Tractatus*, London: Hutchinson University Library.

Antognetti, Paolo and Veljko Milutinovic, (eds.), (1991). *Neural Networks: Concepts, Applications, and Implementations, Volume III*, New Jersey: Prentice Hall.

Arbib, Michael A., and Allen R. Hanson, (eds.), (1987). *Vision, Brain, and Cooperative Computation*, Cambridge: MIT Press.

Ashby, W. Ross (1960). *Design for a Brain*, London: Chapman and Hall.

Ayres, Frank, Jr. (1952). *Theory and Problems of Differential Equations*, New York: Schaum Publishing Co.

Baars, Bernard J. (1998). *A Cognitive Theory of Consciousness*, New York, Oxford: Oxford University Press.

Bacon, Francis (1960). *The New Organon and Related Writings*, F. Anderson, (ed.), New York: Liberal Arts Press.

Bak, P., Tang, C., and Wiesenfeld, K. (1988). "Self-Organized Criticality", *Phys. Rev.*, A, Vol. 38, p. 364.

Bartlett, Frederic C. (1958). *Thinking*, New York: Basic Books.

Baumgartner, T., Burns, Tom R., *et al.* (1976). "Open Systems and Multilevel Processes: Implications for Social Research," in *International Journal of General Systems*, Vol. 3, No. 1, pp. 25-42.

Becker, Gavin De (1997). *The Gift of Fear and Other Survival Signals That Protect Us From Violence*, New York: Dell Publishing.

Becker, S. (1991). "Unsupervised Learning Procedures for Neural Networks," *International Journal of Neural Systems*, Volume 2, pp. 17-33.

Bell, D.J. (1990). *Mathematics of Linear and Nonlinear Systems*, New York, Oxford: Oxford University Press.

Benacerraf, Paul, and Hilary Putnam, (eds.), (1964). *Philosophy of Mathematics*, New Jersey: Prentice-Hall.

Bertalanffy, Ludwig von (1968). *General System Theory: Foundations, Development, Applications,* New York: George Braziller.

Berthoz, Alain, Isabelle Israël, Pierre Georges-François, Renato Grasso, Toshihiro Tsuzuku (1995). "Spatial Memory of Body Linear Displacement: What is Being Stored?" *Science*, American Association for the Advancement of Science, Volume 269, 7 July, pp. 95-98.

Bjorkman, M., P. Juslin, & A. Winman (1993). "Realism of confidence in sensory discrimination: The underconfidence Phenomenon," in *Perception & Psychophysics*, Vol. 54, 1993, pp. 75-81

Black, Max (1964). *A Companion to Wittgenstein's Tractatus*, Ithica, New York: Cornell University Press.

Block, Ned (1995). "On a Confusion About a Function of Consciousness," in *Behavioral and Brain Sciences,* Volume 18, pp. 227-287.

Blum, Adam (1992). *Neural Networks in C++, An Object-Oriented Framework for Building Connectionist Systems,* New York: John Wiley & Sons, Inc.

Blum, L., Shub, M. and S. Smale (1989). "On a Theory of Computation and Complexity over the Real Numbers: NP Completeness, Recursive Functions and Universal Machines," in *The Bulletin of the American Mathematical Society*, Vol. 21. No. 1, July, pp. 1-46.

Blum, L. (1990a). "Lectures on a Theory of Computation and Complexity over the Reals (or an Arbitrary Ring)," in *Lectures in the Sciences of Complexity* II, E. Jen (ed.), Reading, Massachusetts: Addison-Wesley Publishing Co.

Blum, L. (1990b). *A Theory of Computation and Complexity Defined over the Real Numbers*, Technical Report, Berkeley: International Computer Science Institute.

Blum, L. and S. Smale (1990c). *The Gödel Incompleteness Theorem and Decidability Over a Ring*, Technical Report, Berkeley, California: International Computer Science Institute.

Bonevac, Daniel (1991). "Ethical Impressionism: A Response to Braybrooke," in *Social Theory and Practice,* Volume 17, no. 2, Summer, pp. 157-173.

Bower, T.G.R. (1972). "The Visual World of Infants," December, 1966, in *Perception: Mechanisms and Models*, San Francisco: W.H. Freeman and Company, pp. 349-357.

Bradley, M.C. (1969). "Comments and Criticism: How Never to Know What You Mean," in *The Journal of Philosophy,* Vol. LXVI, No. 5, March 13.

Brooks, Rodney and Pattie Maes, (eds), (1994). *Artificial Life IV*, Cambridge: MIT Press.

Bruner, J., Olver, Greenfield, *et al.* (1966). *Studies in Cognitive Growth*, New York: John Wiley & Sons, Inc.

Cantor, Georg (1955). *Contributions to the Founding of the Theory of Transfinite Numbers*, New York: Dover Publications.

Cartwright, Richard L. (1967). "Classes and Attributes," *Noûs* 1, pp. 231-242.

Castañeda, Hector-Ñeri (1967). "Indicators and Quasi-indicators," in *American Philosophical Quarterly,* Vol. 4, pp. 85-100.

Castañeda, Hector-Ñeri (1975). "Individuation and Non-Identity: A New Look," in *American Philosophical Quarterly,* Vol. 12, pp. 131-140.

Castañeda, Hector-Ñeri (1977). "Perception, Belief, and the Structure of Physical Objects and Consciousness," *Synthese,* Vol. 35, pp. 285-351.

Castañeda, Hector-Ñeri (1981). "The Semiotic Profile of Indexical (Experiential) Reference," in *Synthese,* Vol. 49, pp. 275-316.

Castañeda, Hector-Ñeri (1987). "Self-Consciousness, Demonstrative Reference, and the Self-Ascription View of Believing," in *Philosophical Perspectives I, Metaphysics,* James Tomberlin (ed.), Atascadero, California, Ridgeview, pp. 405-459.

Castañeda, Hector-Ñeri (1989). "Direct Reference, The Semantics of Thinking, and Guise Theory: Constructive Reflections on David Kaplan's Theory of Indexical Reference," pp. 105-144, in *Themes from Kaplan,* Joseph Almog, John Perry, Howard Wettstein, (eds.), New York: Oxford University Press.

Castañeda, Hector-Ñeri (1989). "The Reflexivity of Self-Consciousness: Sameness/Identity, Data for Artificial Intelligence, *Philosophical Topics,* Volume XVII, No. 1, Spring, pp. 27-58.

Castañeda, Hector-Ñeri (1990). "Indexicality: The Transparent Subjective Mechanism for Encountering a World," in *Noûs,* Vol. XXIV, No. 5, December, pp. 735-749.

Castañeda, Hector-Ñeri (1990). "Philosophy as a Science and as a Worldview," *The Institution of Philosophy,* Avner Cohen and Marcelo Dascal, (eds.), Bloomington, Indiana: *Noûs* Publications.

Chaitin, G.J. (1966). "On the Length of Programs for Computing Binary Sequences," in *Journal of the Association of Computing Machinery,* Volume 13, pp. 547-569.

Chaitin, G.J. (1974). "Information Theoretical Limitations of Formal Systems," *Journal of the Association of Computing Machinery,* Volume 21, pp. 403-424.

Chaitin, G.J. (1987). *Algorithmic Information Theory,* Cambridge: Cambridge University Press.

Chalmers, David (1996). *The Conscious Mind: In Search of a Fundamental Theory,* New York, Oxford: Oxford University Press.

Cherry, Colin (1957). *On Human Communication,* Cambridge: MIT Press.

Chisholm, Roderick M. (1977). *Theory of Knowledge,* Second Edition, Prentice-Hall, Inc.

Churchland, Paul and Patricia (1990). "Could a Machine Think?" in *Scientific American,* January.

Churchland, P.S. and T.J. Sejnowski (1992). *The Computational Brain,* Cambridge: MIT Press.

Clark, Andy (1989). *Microcognition,* Cambridge: MIT Press.

Cohen, Jack and Ian Stewart (1994). *The Collapse of Chaos: Discovering Simplicity in a Complex World,* New York: Viking, The Penguin Group.

Colombo, John, Jennifer S. Ryther, Janet Frick, Jennifer Gifford (1995). "Visual Pop-out in Infants: Evidence for Preattentive Search in 3- and 4-month-olds," *Psychonomic Bulletin & Review,* Vol 2, number 2, June, 1995, pp. 266-268.

Collishaw, Stephan M., and Graham J. Hole (2000). "Featural and configurational processes in the recognition of faces of different familiarity," in Perception 2000, Volume 29, number 8, pp. 893-909.

Cormen, Thomas H., Charles E. Leiserson, *et al.* (1992). (eds.), *Algorithms,* Cambridge: MIT Press.

Crick, Francis and C. Koch (1990). "Towards a Neurobiological Theory of Consciousness," *Seminars in the Neurosciences,* Vol. 2, pp. 263-275.

Crick, Francis and C. Koch (1992). "The Problem of Consciousness," in *Scientific American,* Volume 267, number 110.

Crick, Francis (1994). *The Astonishing Hypothesis: The Scientific Search for the Soul,* New York: Simon and Schuster.

Culham, Jody C., and Kanwisher, Nancy G. (2001). "Neuroimaging of Cognitive Functions in Human Parietal Cortex, in *Current Opinion in Neurobiology,* Vol 11, 2001, pp. 157-163.

Cutland, N. J. (1980). *Computability, An Introduction to Recursive Function Theory,* Cambridge: Cambridge University Press.

Davenport, J.H. and J. Heintz (1988). "Real Quantifier Elimination is Doubly Exponential," in *Algorithms in Real Algebraic Geometry,* special edition of *The Journal of Symbolic Computation,* Vol. 5, nos. 1 and 2, February, April.

Dawson, Geraldine and K. Fischer, (eds.) (1994). *Human Behavior and the Developing Brain.* New York: Guilford.

Dennett, Daniel (1991). *Consciousness Explained,* New York, Boston: Little, Brown and Company.

Descartes, Rene (1960). *Discourse on Method and Meditations,* translated by Laurence J. Lafleur, Indianapolis: The Bobbs-Merrill Company, Inc.

Devaney, Robert L. (1989). *An Introduction to Chaotic Dynamical Systems,* Second Edition, Reading, Massachusetts: Addison-Wesley Publishing Company.

Devaney, Robert L. (1989). "Dynamics of Simple Maps," in *Proceedings of Symposia in Applied Mathematics: Chaos and Fractals,* Volume 39, American Mathematical Society.

Dreyfus, Hubert L. (1992). *What Computers Still Can't Do: A Critique of Artificial Reason,* Cambridge: MIT Press.

Dubois, Didier and Henri Prade (1991). "Fuzzy Labels, Imprecision and Contextual Dependency: Comments on Milan Zeleny's 'Cognitive Equilibrium: a Knowledge-Based Theory of Fuzziness and Fuzzy Sets,'" in *International Journal of General Systems,* Vol. 19, pp. 383-386.

Edman, Irwin (ed.), (1928). "Theaetetus" in *The Philosophy of Plato,* The Jowett Translation, New York: The Modern Library.

Elsasser, W.M. (1966). *Atom and Organism: A New Approach to Theoretical Biology,* Princeton, New Jersey: Princeton University Press.

Elman, Jeffrey L. (1990). "Finding Structure in Time," *Cognitive Science,* Volume 14, pp. 179-211.

Edman, Irwin (1928). (ed.), *The Philosophy of Plato,* The Jowett Translation, New York: The Modern Library.

Egner, Robert E., and Lester E. Denonn (1961). (eds.), *The Basic Writings of Bertrand Russell: 1903-1959,* New York, London, Tokyo: Simon & Schuster, Inc.

Elsasser, Walter M. (1966). *Atom and Organism: A New Approach to Theoretical Biology,* Princeton, New Jersey: Princeton University Press.

Engel, Andreas K. and Wolf Singer (2001). "Temporal binding and the neural correlates of sensory awareness," in *Trends in Cognitive Sciences,* Vol. 5, no. 1, pp.16-25

Estep, M. (1978a). "The Concept of Understanding," in *SISTM Quarterly,* Vol. 1, no. 3, April.

Estep, M. (1978b). "Toward a SIGGS Characterization of Epistemic Properties of Educational Design," in *Applied General Systems Research: Recent Developments and Trends,* G. Klir (ed.), NATO Conference Series, New York: Plenum Press.

Estep, M. (1978c). "Pragmatics of the SIGGS Theory Model Relative to Pedagogical Epistemological Inquiry," in *Avoiding Social Catastrophes and Maximizing Social Opportunities,* SGSR, Washington, D.C.: American Association for the Advancement of Science.

Estep, M. (1978d). "A SIGGS Information Theoretic Characterization of Qualitative Knowing: Cybernetic and SIGGS Theory Models," in *Sociocybernetics, Vol. 2,* Leiden, Boston, London: Martinus Nijhoff Social Sciences Division.

Estep, M. (1979). "Open Systems Characterizations of Epistemic Properties: Implications for Inquiry in the Human Sciences," in *Improving the Human Condition: Quality and Stability in Social Systems,* Berlin, Heidelberg, New York, London: SGSR and Springer-Verlag.

Estep, M. (1981). "Ways of Qualitative Worldmaking: The Nature of Qualitative Knowing," in *Applied Systems and Cybernetics, Vol. II: Systems Concepts, Models, and Methodology,* G. Lasker (ed.), New York, Pergamon Press.

Estep, M. (1984). "Toward Alternative Methods in Systems Analysis: The Case of Qualitative Knowing," in *Cybernetics and Systems Research, Vol. 2,* Robert Trappl (ed.), Holland: Elsevier Science Publishers B.V. (North-Holland).

Estep, M. (1986). "The Concept of Power and Systems Models for Developing Countries," in *Cybernetics and Systems Research: An International Journal,* Washington, D.C.: Hemisphere Publishing Corp. of Harper and Row.

Estep, M. (1987). "Systems Analysis and Power," in *Problems of Constancy and Change: The Complementarity of Systems Approaches to Complexity,* Budapest: Hungarian Academy of Sciences.

Estep, M. (1992). "On Models and Retroductive Inference," in *Cybernetics and Systems Research, Vol. 1,* Robert Trappl (ed.), Singapore, New Jersey, London, Hong Kong: World Scientific.

Estep, M. (1993). "On Qualitative Logical and Epistemological Aspects of Fuzzy Set Theory and Test-Score Semantics: Indexicality and Natural Language Discourse," in *First European Congress on Fuzzy and Intelligent Technologies Proceedings, Vol. 2,* Hans-Jürgen Zimmerman, (ed.), Aachen, Germany: Verlag der Augustinus Buchhandlung.

Estep, M. (1996). "Critique of James' Neutral Monism: Consequences for the New Science of Consciousness," abstract in *Journal of Consciousness Studies*, Imprint Academic.

Estep, M. (1998). "What Gödel Said: On the Non-algorithmic Nature of the Second Theorem," in *Systems Research, EMCSR 1998*, Vienna: Austrian Society for Cybernetic Studies.

Estep, M. (1999). "Teaching the Logical Paradoxes: On Mathematical Insight or Is there a Non-algorithmic Element in Gödel's Second Theorem?" In *Abstracts of Papers Presented to the American Mathematical Society,* Volume 20, Number 1 (Issue 115), January, 1999.

Estep, M. (in progress). *On Philosophical Foundations of Epistemology: The Architecture of Intelligence, and Issues of Decidability and Complexity.*

Eubank, Stephen, and Doyne Farmer (1990). "An Introduction to Chaos and Prediction," in Erica Jen (ed.), *1989 Lectures in Complex Systems*, Santa Fe Institute Studies in the Sciences of Complexity, Reading, Massachusetts: Addison-Wesley Publishing Company.

Familant, M.E., and M.C. Detweiler (1993). "Iconic Reference: Evolving Perspectives and an Organizing Framework," in *International Journal of Man-Machine Studies* 39, pp. 705-728.

Farmer, J. Doyne (1990). "A Rosetta Stone for Connectionism" in *Physica D*, Volume 42, pp. 153-187.

Feferman, Solomon, J.W. Dawson, Jr., Stephen C. Kleene, *et al.* (1986). *Kurt Gödel Collected Works,* Volumes I, New York, Oxford: Oxford University Press.

Feferman, Solomon, J.W. Dawson, Jr., Stephen C. Kleene, *et al.* (1990). *Kurt Gödel Collected Works,* Volume II, New York, Oxford: Oxford University Press.

Fine, Kit (1989). "The Problem of *De Re* Modality," in *Themes From Kaplan*, Joseph Almog, John Perry, and Howard Wettstein, (eds.), New York, Oxford: Oxford University Press.

Forrest, Stephanie and John H. Miller (1991). "Emergent Behavior in Classifier Systems," in *Emergent Computation*, (Stephanie Forrest, ed.), Cambridge: MIT Press, pp. 213-227.

Foster, Lawrence, and J.W. Swanson (1970). *Experience and Theory,* Amherst: University of Massachusetts Press.

Freeman, James A. (1994). *Simulating Neural Networks with Mathematics*, Reading, Massachusetts: Addison-Wesley Publishing Company.

Frege, Gottlob (1892). "Über Begriff und Gegenstand" in *Vierteljahrsschrift für wissenschaftliche Philosophie*, 16, pp. 192-205;

Frege, Gottlob (1952). "Uber Sinn und Bedeutung," (On Sense and Reference), in Geach, Peter, and Max Black, (eds.), *Translations From the Philosophical Writings of Gottlob Frege*, New York, Oxford: Oxford University Press.

Frith, Chris, Richard Perry and Erik Lumer (1999). "The neural correlates of conscious experience: an experimental framework," *Trends in Cognitive Sciences*, Vol. 3, no. 3, pp. 105-114.

Fritzke, Bernd (1993). *Growing Cell Structures--A Self-Organizing Network for Unsupervised and Supervised Learning*, Technical Report, Berkeley: California, International Computer Science Institute.

Fritzke, Bernd (1993). *Kohonen Feature Maps and Growing Cell Structures--A Performance Comparison*, Berkeley, California: Technical Report, International Computer Science Institute.

Gardner, Howard (1982). *Art, Mind, and Brain*, New York: Basic Books.

Gardner, Howard (1985). *The Mind's New Science*, New York: Basic Books.

Gardner, Howard (1993). *Frames of Mind: The Theory of Multiple Intelligences*, New York: Basic Books.
Gärdenfors, Peter (1988). *Knowledge in Flux: Modeling the Dynamics of Epistemic States,* Cambridge: MIT Press.
Geach, Peter (1971). *Mental Acts, Their Content and Their Objects,* New York: Humanities Press.
Gödel, Kurt (1964a). "Russell's Mathematical Logic," in Benacerraf, Paul, and Hilary Putnam (eds.), *Philosophy of Mathematics,* New Jersey: Prentice-Hall.
Gödel, Kurt (1964b). "What is Cantor's Continuum Problem?" in Benacerraf, Paul, and Hilary Putnam (eds.), *Philosophy of Mathematics,* New Jersey: Prentice-Hall.
Gomi, Hiroaki and Mitsuo Kawato (1996). "Equilibrium-Point Control Hypothesis Examined by Measured Arm Stiffness During Multijoint Movement," in *Science*, American Association for the Advancement of Science, Volume 272, 5 April, pp. 117-120.
Goodman, Nelson (1973). *Fact, Fiction, and Forecast*, Bobbs-Merrill Publishing Company.
Graves, Katz, *et al.* (1973). "Tacit Knowledge," in *The Journal of Philosophy*, Vol. LXX, No. 11, June 7.
Grimaldi, Ralph (1994). *Discrete and Combinatorial Mathematics*, Third Edition, Reading, Massachusetts: Addison-Wesley Publishing Company.
Grimson, W. Eric and Ramesh S. Patil (1987). (eds.), *AI in the 1980s and Beyond: An MIT Survey,* Cambridge: MIT Press.
Gupta, Dinkar (2001). "Computer Gesture Recognition: Using the Constellation Method," in *Caltech Undergraduate Research Journal*, Vol. 1, April.
Gurwitsch, Aron (1963). "On the Conceptual Consciousness," in *The Modeling of Mind,* Kenneth M. Sayre and Frederick J. Crosson, (eds.), South Bend: Notre Dame University Press.
Hadamard, Jacques (1945). *The Psychology of Invention in the Mathematical Field,* Princeton: Princeton University Press.
Hanson, Norwood Russell (1972). *Patterns of Discovery,* Cambridge: Cambridge University Press.
Harary, F., (1969), *Graph Theory*, Reading, Massachusetts: Addison-Wesley.
Harmon, Gilbert Harmon (1977). *Thought*, Princeton, New Jersey: Princeton University Press.
Harmon, Leon D. (1973). "Recognition of Faces," in *Scientific American,* November.
Hartley, Ralph V. L., (1928). "Transmission of Information," in *The Bell Systems Technical Journal,* Vol. 7, pp. 535-563.
Hartshorne, Charles and Weiss, Paul, (eds.), (1958). *The Collected Papers of Charles Sanders Peirce,* Vols. I-VI, Cambridge: Harvard University Press,.
Hausman, Alan, and Tom Foster (1977). "Is Everything a Class?" in *Philosophical Studies*, 32, pp. 371-376.
Haykin, Simon (1994). *Neural Networks: A Comprehensive Foundation*, New York: Macmillan College Publishing Company.
Healey, C. G. (1993). "Visualization of Multivariate Data Using Preattentive Processing." Masters Thesis, Department of Computer Science, University of British Columbia.
Hebb, D.O. (1949). *The Organization of Behavior: A Neuropsychological Theory*, New York: John Wiley and Sons, Inc.
Heijenoort, Jean van (1967). (ed.) *From Frege to Gödel,* Cambridge: Harvard University Press.
Hempel, Carl G. (1965). *Aspects of Scientific Explanation*, New York, London: The Free Press.
Hernegger, R. (1995). *Wahrnehmung und Bewusstsein: Ein Diskussionsbeitrag zur Neuropsychologie*, Heidelburg: Spektrum.
Higashi, M., and G. Klir (1982). "Measures of Uncertainty and Information Based on Possibility Distributions," in *International Journal of General Systems,* 8, Washington, D.C.: SGSR.

Higashi, M., and G. Klir (1983). "On the Notion of Distance Representing Information Closeness," *International Journal of General Systems*, 9,Washington, D.C.: ISGSR.

Hinton, G.E. (1981). "Shape Representation in Parallel Systems," *Proceedings of the 7th International Joint Conference on Artificial Intelligence,* Vancouver, British Columbia.

Hinton, G.E. and T.J. Sejnowski (1986). "Learning and Relearning in Boltzmann Machines," in *Parallel Distributed Processing: Explorations in Microstructure of Cognition* (D.E.Rumelhart and J.L. McClelland, eds.), Cambridge: MIT Press.

Hinton, G.E., Peter Dayan, Brendan Frey, Radford Neal (1995). "The 'Wake-Sleep' Algorithm for Unsupervised Neural Networks" in *Science*, American Association for the Advancement of Science, Volume 268, Number 5214, 26 May, pp. 1158-1161.

Hochberg, Herbert (1978). *Thought, Fact, and Reference: The Origins and Ontology of Logical Atomism*, Minneapolis: University of Minnesota.

Holland, John (1975). *Adaptation in Natural and Artificial Systems*, Ann Arbor: University of Michigan Press.

Holland, John H., Keith J. Holyoak, Richard Nisbett, and Paul Thagard (1986). *Induction: Processes of Inference, Learning, and Discovery*, Cambridge: MIT Press.

Jackson, Peter, Reichgelt, Han, and Frank van Harmelen (1989). *Logic-Based Knowledge Representation,* Cambridge: MIT Press.

James, William (1884). "Some Omissions of Introspective Psychology," *Mind*, 9, January, 1884, pp. 1-26.

James, William (1890). *The Principles of Psychology, Volumes I and II*, London: Macmillan.

James, William (1976). *Essays in Radical Empiricism*, Cambridge: Harvard University Press.

Jen, Erica (1989). (ed.), *1989 Lectures in Complex Systems*, Santa Fe Institute Studies in the Sciences of Complexity, Reading, Massachusetts: Addison-Wesley Publishing Company.

Johnson-Laird, P.N. (1983). *Mental Models*, Cambridge: Harvard University Press.

Jordan, M.I. (1986). "An Introduction to Linear Algebra in Parallel Distributed Processing," in *Parallel Distributed Processing: Explorations in the Microstructure of Cognition, Volume 1: Foundations,* David E. Rumelhart, James L. McClelland, PDP Research Group, Cambridge: MIT Press.

Jourdain, Philip Edward Bertrand (1912). "The Development of the Theories of Mathematical Logic and the Principles of Mathematics," in *The Quarterly Journal of Pure and Applied Mathematics,* 43, pp. 219-314.

Jusczyk, Peter W. (1997). *The Discovery of Spoken Language*, Cambridge: MIT Press.

Kandel, E.R. and J.H. Schwartz (1991). *Principles of Neural Science*, 3rd edition, New York: Elsevier.

Kant, Immanuel (1783). *Prolegomena to any Future Metaphysics that will be able to present itself As a Science,* Riga: Johann Friedrich Hartknoch.

Kant, Immanuel (1929). *Critique of Pure Reason,* Norman Kemp Smith, translator, Toronto: Macmillan & Co.

Kaplan, David (1989). "Demonstratives," in *Themes From Kaplan*, Joseph Almog, John Perry, Howard Wettstein, (eds.), New York, Oxford: Oxford University Press.

Kauffman, Stuart A. (1990). "Requirements for Evolvability in Complex Systems: Orderly Dynamics and Frozen Components," in *Physica D* Volume 42, , pp. 135-152.

Kauffman, Stuart A. (1991). "Antichaos and Adaptation," in *Scientific American*, Volume 265, Number 2, August, pp. 78-84.

Kauffman, Stuart A. (1993). *The Origins of Order: Self-Organization and Selection in Evolution*, New York, Oxford: Oxford University Press.

Kauffman, Stuart A. (1995). *At Home in the Universe: The Search for the Laws of Self-Organization and Complexity*, New York, Oxford: Oxford University Press.

Keen, L. (1989). "Julia Sets," in *Proceedings of Symposia in Applied Mathematics: Chaos and Fractals,* Volume 39, American Mathematical Society.

Kellert, Stephen H., Mark A. Stone, Arthur Fine (1990). "Models, Chaos, and Goodness of Fit," in *Philosophical Topics*, Vol. 18, No. 2, Fall.

Kerlinger, Fred (1973). *Foundations of Behavioral Science*, 2nd edition, New York: Holt, Rinehart and Winston, Inc.

Kessen, William (1965). *Child*, New York: John Wiley & Sons, Inc.

Kohonen, T. (1982). "Self-organized Formation of Topologically Correct Feature Maps," *Biological Cybernetics*, Volume 43, , pp. 59-69.

Kohonen, T. (1988). "An Introduction to Neural Computing," in *Neural Networks*, Volume 1, pp. 3-16.

Kohonen, T. (1990). "The Self-organizing Map," *Proceedings of the IEEE*, 78, pp. 1464-1480.

Kornblith, Hilary (1994). *Naturalizing Epistemology* 2nd Edition, Cambridge: MIT Press

Kornblith, Hilary (1999). "In Defense of a Naturalized Epistemology," in *The Blackwell Guide to Epistemology*, John Greco and Ernest Sosa (eds.), Oxford: Basil Blackwell.

Kosslyn, Stephen (1996). *Image and Brain*, Cambridge: MIT Press.

Kosslyn, Stephen (2002). "*Visual Mental Images in the Brain: How Low Do They Go*," presented at a meeting of the American Association for the Advancement of Science on the Cognitive Neuroscience of Mental Imagery, February.

Kripke, Saul (1972). *Naming and Necessity*, Cambridge: Harvard University Press.

Kuhn, Thomas S. (1970). *The Structure of Scientific Revolutions*, 2nd Edition, International Encyclopedia of Unified Science, Chicago: University of Chicago Press.

Kunimoto, C., *et al.* (2001). "Confidence and Accuracy in Near-Threshold Discrimination Responses," in *Consciousness and Cognition*, Vol. 10, no. 3, pp. 294-340.

Lakoff, George and Mark Johnson (1998). *Philosophy in the Flesh: The Embodied Mind and Its Challenge to Western Thought*, New York: Basic Books.

Langton, Christopher G., (ed.), (1989). *Artificial Life,* Santa Fe Institute Studies in the Sciences of Complexity, Vol. 6, Redwood City, California: Addison-Wesley Publishing Co.

Lay, Steven R. (1990). *Analysis With An Introduction to Proof*, Second Edition, New Jersey: Prentice-Hall.

Lehrer, Keith, and Thomas Paxson, Jr. (1968). "Knowledge: Undefeated Justified True Belief," *Journal of Philosophy,* Vol. LXVI, No. 8, April.

Lehrer, Keith (1974). *Knowledge,* New York, Oxford: Oxford University Press.

Lehrer, Keith (1980). "Knowledge," in Bogdan, R. J., (ed.), *Keith Lehrer,* D. Reidel Publishing Company.

Lehrer, Keith (1983). "Coherence and Indexicality in Knowledge," in James E. Tomberlin, (ed.), *Agent, Language, and the Structure of the World: Essays Presented to Hector-Neri Castañeda With His Replies*, Atascadero, California: Ridgeview Publishing Co.

Lenat, Douglas B. (1995). "Artificial Intelligence: A Critical Storehouse of Commonsense Knowledge is Now Taking Shape," in *Scientific American*, September, pp. 80-82.

Lewis, David (1979). "Attitudes *De Dicto* and *De Se*," in *Philosophical Review,* Vol. 88, pp. 513-543.

Li, Fei Fei, and Rufin Van Rullen, Christof Koch & Pietro Perona (2002). "Rapid natural scene categorization in the near absence of Awareness" in *Proc Nat Acad Sci,* vol 99, July.

Libet, B. (1973). "Electrical Stimulation of Cortex in Human Subjects, and Conscious Memory Aspects," in A. Iggo (eds.), *Handbook of Sensory Physiology*, Vol. II, Berlin, Heidelberg, New York: Springer-Verlag.

Lipton, Richard J. (1995). "DNA Solution for Hard Computational Problems," *Science,* American Association for the Advancement of Science, Vol. 268, 28 April, pp. 542-545.

Livingstone, M., and Hubel, D., "Segregation of Form, Color, Movement, and Depth: Anatomy, Physiology, and Perception," in *Science*, Vol. 240, 1988, pp. 740-749.

Loux, Michael, J. (1970). *Universals and Particulars, Readings in Ontology*, New York: Anchor Books, Doubleday and Company, Inc.

Luria, Alexander R. (1968). *The Mind of the Mnemonist*, Cambridge: Harvard University Press.
Maccia, George (1987). "Genetic Epistemology of Intelligent, Natural Systems," in *Systems Research*, Volume 3, 1987.
Maccia, George (1989). *Genetic Epistemology of Intelligent Systems: Propositional, Procedural, and Performative Intelligence,* presented at Hangzhou University, Hangzhou, Zhejiang Province, The People's Republic of China.
Mach, Ernst (1959). *Analysis of the Sensations,* New York: Dover Publications.
Mager, Robert F. (1975). *Preparing Instructional Objectives*, 2nd edition, Belmont, California: Feron Publishing Company.
Malcolm, Norman (1958). *Ludwig Wittgenstein: A Memoir,* Oxford: Oxford University Press.
Mandelbrot, Benoit (1983). *The Fractal Geometry of Nature*, New York: W. H. Freeman and Company.
Marsh, Robert C. (1956). (ed.), *Bertrand Russell: Logic and Knowledge Essays 1901-1950,* New York: Capricorn Books.
Martin, Edwin (1973). "The Intentionality of Observation," in *Canadian Journal of Philosophy,* Volume III, Number 1, September, pp. 121-129.
McClelland, James L., David E. Rumelhart and the PDP Research Group (1986). *Parallel Distributed Processing,* Volumes 1 and 2, Cambridge: MIT Press.
Meinong, Alexius (1899). "Über Gegenstände höherer Ordnung und deren Verhältniss zur inneren Wahrnehmung," in *Zeitschrift für Psychologie des Sinnesorgane*, 21, pp. 182-272.
Mitra, Sushmita, Sankar K. Pal (1994). "Self-Organizing Neural Network as A Fuzzy Classifier," in *IEEE Transactions on Systems, Man, and Cybernetics*, Volume 24, No. 3, March, pp. 385-398.
Moss, Frank and Kurt Wiesenfeld (1995). "The Benefits of Background Noise," in *Scientific American,* Volume 273, Number 2, August, pp. 66-69.
Näätanen, Risto, Mari Tervaniemi, Elyse Sussman, Petri Paavilinen and Istvan Winkler (2001). "'Primitive Intelligence' in the Auditory Cortex," *Trends in Neurosciences*, Vol 24, number 5, pp. 283-288
Nagel, Ernest, and James Newman (1958). *Gödel's Proof,* New York: New York University Press.
Nicolelis, Miguel A., Luiz A. Baccala, Rick C.S. Lin, John K. Chapin (1995). "Sensorimotor Encoding by Synchronous Neural Ensemble Activity at Multiple Levels of the Somatosensory System," *Science*, American Association for the Advancement of Science, Volume 268, 2 June, pp. 1353-1358.
Pearson, K. (1976). "The Control of Walking," in *Scientific American*, Vol. 235, pp. 72-86.
Perry, John (1979). "The Problem of the Essential Indexical," in *NOÛS*, 13, pp. 3-21.
Penrose, Roger (1974). "The Role of Aesthetics in Pure and Applied Mathematical Research," in the *Bulletin of the Institute of Mathematics and Its Applications,* July/August, pp. 266-271.
Penrose, Roger (1989). *The Emperor's New Mind*, New York, Oxford: Oxford University Press.
Penrose, Roger (1994). *Shadows of the Mind*, New York, Oxford: Oxford University Press.
Perry, John (1979). "The Problem of the Essential Indexical," in *NOÛS*, 13, 1979, pp. 3-21.
Piaget, J. and B. Inhelder (1956). *The Child's Conception of Space*, London: Routledge & Kegan Paul.
Pojman, Louis P. (1995). *What Can We Know, An Introduction to the Theory of Knowledge*, Belmont: Wadsworth Publishing Co.
Polanyi, Michael (1969). "The Unaccountable Element in Science," in *Knowing and Being,* Chicago: University of Chicago Press.
Polanyi, Michael (1967). *The Tacit Dimension*, New York: Doubleday & Company, Inc.

Polanyi, Michael (1969). *Knowing and Being*, Marjorie Grene, (ed.), Chicago: The University of Chicago Press.

Popper, Karl (1972). *Objective Knowledge*, Oxford: Clarendon Press.

Popper, Karl and Eccles, John C. (1977). *The Self and Its Brain,* New York, Heidelberg, London: Springer-Verlag.

Quine, W.V.O. (1951). "Two Dogmas of Empiricism," in *The Philosophical Review*, Vol. 60, 1951, pp. 20-43.

Quine, W.V.O. (1953). *From a Logical Point of View,* Cambridge: Harvard University Press.

Quine, W.V.O. (1960). *Word and Object,* Cambridge: MIT Press.

Quine, W.V.O. (1963). *Set Theory and Its Logic,* Cambridge: Harvard University Press.

Quine, W.V.O. (1969). *Ontological Relativity and Other Essays*, New York: Columbia University Press.

Quine, W.V.O. (1970). "Grades of Theoreticity," in *Experience and Theory*, Foster and Swanson (eds.), Amherst: University of Massachusetts Press, pp. 4-7.

Quine, W.V.O. and J.S. Ullian (1978). *Web of Belief*, Second Edition, New York: Random House.

Quine, W. V. O. (1981). *Theories and Things,* Cambridge: Harvard University Press.

Quine, W.V.O. (1990). "Norms and Aims," in *The Pursuit of Truth*, Cambridge: Harvard University Press.

Rizzolatti, Giacomo and Michael A. Arbib (1998). "Language Within Our Grasp," in *Trends in Neuroscience,* Volume 21, number 5, 1998, pp. 188-194.

Renegar, J. (1988). "A Faster PSPACE Algorithm for Deciding the Existential Theory of the Reals," in *Proceedings of the 29th Annual Symposium of Computer Science*, October, IEEE Computer Society Press.

Repp, Bruno (2001). "Phase Correction, Phase Resetting, and Phase Shifts After Subliminal Timing Perturbations in Sensorimotor Synchronization," *Journal of Experimental Psychology: Human Perception and Performance*, American Psychological Association, Vol. 27, Number 3, June, 2001.

Rosen, R. (1970). *Dynamical System Theory in Biology, Vol. I: Stability Theory and Its Applications*, New York: John Wiley & Sons, Inc.

Rowland, Todd (1999). "Manifold," in *Eric Weisstein's Math World*, Chicago: Stephen Wolfram Research, Inc.

Rucker, Rudy (1982). *Infinity and the Mind: The Science and Philosophy of the Infinite*, New York: Bantam Books.

Russell, Bertrand (1903). *Principles of Mathematics*, New York: W. W. Norton & Company, Inc.

Russell, Bertrand (1911-1912). "On the Relations of Universals and Particulars," *Proceedings of the Aristotelian Society.*

Russell, Bertrand (1912). *The Problems of Philosophy*, London: Thornton Butterworth Ltd.

Russell, Bertrand (1914)."Preliminary Description of Experience," in *The Monist,* 24, (January), pp. 1-16.

Russell, Bertrand (1915). *Our Knowledge of the External World,* Chicago: Open Court Publishing Co.

Russell, Bertrand (1918). *Mysticism and Logic,* London: Penguin Books.

Russell, Bertrand (1921). *The Analysis of Mind*, London: George Allen & Unwin Ltd.

Russell, Bertrand (1927). *An Outline of Philosophy,* London: Allen & Unwin.

Russell, Bertrand (1940). "Language and Metaphysics," in *An Inquiry into Meaning and Truth*, London: George Allen and Unwin Ltd.

Russell, Bertrand (1948). *Human Knowledge*, New York: Simon and Schuster.

Russell, Bertrand (1984). *Theory of Knowledge: The 1913 Manuscript*, Elizabeth Ramsden Eames, (ed.), London and New York: Allen & Unwin.

Ryle, Gilbert (1949). *The Concept of Mind*, New York, London: Barnes and Noble Books.
Saaty, Thomas L. and Joseph Bram (1964) (eds.), *Nonlinear Mathematics,* New York: Dover Publications.
Sayre, Kenneth and Frederick Crosson (1963). *The Modeling of Mind*, South Bend, Indiana: Notre Dame University Press.
Scheffler, Israel (1965). *Conditions of Knowledge: An Introduction to Epistemology and Education*, Glenview, Illinois: Scott, Foresman and Company.
Schilpp, Paul Arthur (ed.), (1946). *The Philosophy of Bertrand Russell,* Illinois: The Library of Living Philosophers, Inc.
Scott, Alwyn (1995). *Stairway to the Mind: The Controversial New Science of Consciousness*, New York: Springer-Verlag.
Searle, John (1967). "Proper Names and Descriptions," *Encyclopedia of Philosophy*, Paul Edwards (ed.), Volume 6, New York: Macmillan Publishing Company.
Searle, John (1992). *The Rediscovery of the Mind*, Cambridge: MIT Press.
Searle, John (1995). "The Mystery of Consciousness," in *The New York Review of Books*, November and December, New York: The New York Times, Inc.
Sejnowski, Terrance, J. and Geoffrey E. Hinton (1987; 1990). "Separating Figure from Ground with a Boltzmann Machine," in Michael Arbib and Allen Hanson (eds.), *Vision, Brain, and Cooperative Computation*, Cambridge: MIT Press.
Shafer, Glen (1976). A Mathematical Theory of Evidence, Princeton: Princeton University Press.
Shannon, C., and Warren Weaver (1949). *The Mathematical Theory of Communication*, Urbana: University of Illinois Press.
Singh, Simon (1997). *Fermat's Enigma: The Epic Quest to Solve the World's Greatest Mathematical Problem,* New York: Walker & Co., 1997.
Sluga, Hans (1980). *Gottlob Frege*, London: Routledge and Kegan Paul.
Steels, Luc (1993). "The Artificial Life Roots of Artificial Intelligence, *Artificial Life*, Vol. 1, Number 1/2, Cambridge: MIT Press.
Stein, Daniel L. (1988). *Lectures in the Sciences of Complexity*, Santa Fe Institute Studies in the Sciences of Complexity, Addison-Wesley Publishing Company.
Steiner, E. (1976). "Logical and Conceptual Analytic Techniques for Educational Researchers," in *Proceedings of the American Educational Research Association*, San Francisco, Washington, D.C.: American Educational Research Association.
Steiner, Elizabeth (1988). *Methodology of Theory Building*, Sydney, Australia: Educology Research Associates.
Stewart, Ian (1995). *Nature's Numbers*, New York: Basic Books.
Stich, Stephen, and Richard Nisbett (1980). "Justification and the Psychology of Human Reasoning," in *Philosophy of Science*, Vol. 47, pp. 188-202.
Stich, Stephen (1990). *The Fragmentation of Reason*, Cambridge, Massachusetts: MIT Press
Stix, Gary (1995). "Boot Camp for Surgeons," *Scientific American*, September, p. 24
Stout, George Frederick (1901). *A Manual of Psychology*, 2nd edition, London: University Tutorial Press.
Tanenhaus, Michael K., Michael J. Spivey-Knowlton, Kathleen Eberhard, Julie Sedivy (1995). "Integration of Visual and Linguistic Information in Spoken Language Comprehension," *Science*, American Association for the Advancement of Science, Volume 268, 16 June, pp. 1632-1634.
Tarski, Alfred (1951). *A Decision Method for Elementary Algebra and Geometry,* 2nd revised edition, Berkeley, California: University of California Press.
Tomberlin, James (ed.), (1983). *Agent, Language, and the Structure of the World: Essays Presented to Hector-Neri Castañeda with His Replies,* Atascadero, California: Ridgeview Publishing Company and Indianapolis: Hackett Publishing Co.

Turing, A. M. (1937). "On Computable Numbers With An Application to the *Entscheidungsproblem*," in *Proceedings of the London Mathematical Society*, Volume 42, pp. 230-265.

Vikhanski, Luba (2001). *In Search of the Lost Cord*, Washington, D.C.: Joseph Henry Press.

Vinod, V.V., Santanu Chaudhury, J. Mukherjee, and S. Ghose (1994). "A Connectionist Approach for Clustering with Applications in Image Analysis" in *IEEE Transactions on Systems, Man, and Cybernetics,* Vol. 24, No. 3, March, pp. 365-383.

Von Bertalanffy, Ludwig (1968). *General System Theory*, New York: George Braziller.

Webster's Encyclopedic Unabridged Dictionary, New York: Portland House, 1989.

Weiskrantz, Lawrence (1997). *Consciousness Lost and Found*, New York, Oxford: Oxford University Press.

Wittgenstein, Ludwig (1922). *Tractatus Logico-Philosophicus*, C.K. Ogden, translator, London: Routledge & Kegan Paul Ltd.

Wittgenstein, Ludwig (1953). *Philosophical Investigations*, Third Edition, G.E.M. Anscombe, translator, New York: Macmillan Publishing Co., Inc.

Wittgenstein, Ludwig (1969). *Über Gewissheit: On Certainty,* G.E.M. Anscombe and G. H. von Wright, (eds.), New York, London: Harper and Row.

Wolfe, J. M. (1996). "Visual Search," in H. Pashler (ed.), *Attention*, London: University College London Press.

Wolfe, J. M. and Sara C. Bennett (1997). "Preattentive Object Files: Shapeless Bundles of Basic Features," in *Vision Research*, Vol. 37, Issue 1, January.

Wolfram, Stephen (1984). "Computer Software in Science and Mathematics," in *Scientific American*, September, pp. 188-203.

Wolfram, Stephen (ed.), (1986). *Theory and Applications in Cellular Automata,* Singapore: World Scientific.

Wolpaw, J. R. (1997). "The Complex Structure of Simple Memory," in *Trends in Neurosciences*, Vol. 20, pp. 588-594.

Wright, C. (1983). *Frege's Conception of Numbers as Objects*, Aberdeen: Aberdeen University Press.

Wright, S. (1931). "Evolution in Mendelian Populations," *Genetics*, Vol. 16, number 97.

Wright, S. (1932). "The Roles of Mutation, Inbreeding, Crossbreeding and Selection in Evolution," *Proceedings of the Sixth International Congress in Genetics*, Vol. 1, number 356.

Zadeh, L.A. (1971). "Fuzzy Languages and Their Relation to Human and Machine Intelligence," in *Proceedings of the Conference on Man and Computer,* Bordeaux, France, Memorandum M-302, Electronics Research Laboratory, University of California at Berkeley.

Zadeh, L.A. (1973). "Outline of a New Approach to the Analysis of Complex Systems and Decision Processes,"*IEEE Transactions on Systems, Man, and Cybernetics,* Vol. SMC-3, No. 1.

Zadeh, L.A. (1977a). *A Theory of Approximate Reasoning,* Memorandum No. UCB/ERL M77/58, Electronics Research Lab., College of Engineering, University of California at Berkeley.

Zadeh, L.A. (1977b). *PRUF-A Meaning Representation Language,* Memorandum No. ERL-M77/61, Electronics Research Lab., College of Engineering, University of California at Berkeley.

Zadeh, L.A. (1978). "PRUF-A Meaning Representation Language for Natural Languages," *International Journal of Man-Machine Studies*, Vol. 10, pp. 395-460.

Zadeh, L.A. (1981). "Test-Score Semantics for Natural Languages and Meaning Representation Via PRUF," in *Empirical Semantics,* Burghard B. Rieger (ed.), Studienverlag Dr. N. Brockmeyer, Bochum.

Zadeh, L.A. (1983). "A Fuzzy Set-Theoretic Approach to the Compositionality of Meaning: Propositions, Dispositions and Canonical Forms," Memorandum No. UCB/ ERL M83/24, 4 April.

Zadeh, L.A., *et al.* (1990). (eds.), *Uncertainty in Knowledge Bases,* Berlin, Heidelberg, New York, London: Springer-Verlag.

Zalta, E. (1999). "Natural Numbers and Natural Cardinals as Abstract Objects: A Partial Reconstruction of Frege's *Grundgesetze* in Object Theory," in *Journal of Philosophical Logic*, Vol. 28, number 6, pp. 619-660.

INDEX

MIX
Papier aus verantwortungsvollen Quellen
Paper from responsible sources
FSC® C105338

FSC
www.fsc.org

If you have any concerns about our products,
you can contact us on
ProductSafety@springernature.com

In case Publisher is established outside the EU,
the EU authorized representative is:
Springer Nature Customer Service Center GmbH
Europaplatz 3, 69115 Heidelberg, Germany

Printed by Libri Plureos GmbH
in Hamburg, Germany